Semiconductor Devices
for Electronic Tuners

Japanese Technology Reviews

A series of tracts that examines the status of and future prospects for Japanese technology in five fields: **electronics, computers and communications, manufacturing engineering, new materials, and biotechnology.**

Editor in Chief

Toshiaki Ikoma, *University of Tokyo*

Section Editors

Electronics

MMIC—Monolithic Microwave Integrated Circuits *(Volume 2)*
Yasuo Mitsui

Bulk Crystal Growth Technology *(Volume 4)*
Shin-ichi Akai, Keiichiro Fujita, Masamichi Yokogawa, Mikio Morioka and Kazuhisa Matsumoto

Semiconductor Heterostructure Devices *(Volume 8)*
Masayuki Abe and Naoki Yokoyama

Development of Optical Fibers in Japan *(Volume 11)*
Hiroshi Murata

High-Performance BiCMOS Technology and Its Applications to VLSIs *(Volume 12)*
Ikuro Masuda and Hideo Maejima

Semiconductor Devices for Electronic Tuners *(Volume 13)*
Seiichi Watanabe

Japanese Technology Reviews

Computers and Communications

Machine Vision: A Practical Technology for Advanced Image Processing *(Volume 10)*
Masakazu Ejiri

Manufacturing Engineering

Automobile Electronics *(Volume 1)*
Shoichi Washino

Steel Industry I: Manufacturing System *(Volume 6)*
Tadao Kawaguchi and Kenji Sugiyama

Steel Industry II: Control System *(Volume 7)*
Tadao Kawaguchi and Takatsugu Ueyama

Networking in Japanese Factory Automation *(Volume 9)*
Koichi Kishimoto, Hiroshi Tanaka,
Yoshio Sashida and Yasuhisa Shiobara

Biotechnology

Production of Nucleotides and Nucleosides by Fermentation *(Volume 3)*
Sadao Teshiba and Akira Furuya

Recent Progress in Microbial Production of Amino Acids *(Volume 5)*
Hitoshi Enei, Kenzo Yokozeki and Kunihiko Akashi

This book is part of a series. The publisher will accept continuation orders which may be cancelled at any time and which provide for automatic billing and shipping of each title in the series upon publication. Please write for details.

Semiconductor Devices for Electronic Tuners

by
Seiichi Watanabe
Sony Research Center
Yokohama, Japan

Gordon and Breach Science Publishers

New York • Philadelphia • London • Paris • Montreux • Tokyo • Melbourne

Gordon and Breach Science Publishers

Post Office Box 786
Cooper Station
New York, New York 10276
United States of America

5301 Tacony Street, Drawer 330
Philadelphia, Pennsylvania 19137
United States of America

Post Office Box 197
London WC2E 9PX
United Kingdom

58, rue Lhomond
75005 Paris
France

Post Office Box 161
1820 Montreux 2
Switzerland

3-14-9, Okubo
Shinjuku-ku, Tokyo 169
Japan

Private Bag 8
Camberwell, Victoria 3124
Australia

Library of Congress Cataloging-in-Publication Data

Watanabe, Seiichi.
 Semiconductor devices for electronic tuners / by Seiichi Watanabe.
  p. cm.—(Japanese technology reviews, ISSN 0898-5693 : vol. 13)
 ISBN 2-88124-475-0
 1. Television—Tuners. 2. Semiconductors. I. Title.
 II. Series.
 TK6655.T8W38 1991
 621.388—dc20   90-14074
                                                                  CIP

To
Akinori Takasaki
for his encouragement
throughout my business career

Contents

Contents

Preface to the Series

Modern technology has a great impact on both industry and society. New technology is first created by pioneering work in science. Eventually, a major industry is born, and it grows to have an impact on society in general. International cooperation in science and technology is necessary and desirable as a matter of public policy. As development progresses, international cooperation changes to international competition, and competition further accelerates technological progress.

Japan is in a very competitive position relative to other developed countries in many high technology fields. In some fields, Japan is in a leading position; for example, manufacturing technology and microelectronics, especially semiconductor LSIs and optoelectronic devices. Japanese industries lead in the application of new materials such as composites and fine ceramics, although many of these new materials were first developed in the United States and Europe. The United States, Europe, and Japan are working intensively, both competitively and cooperatively, on the research and development of high-critical-temperature superconductors. Computers and communications are now a combined field that plays a key role in the present and future of human society. In the next century, biotechnology will grow, and it may become a major segment of industry. While Japan does not play a major role in all areas of biotechnology, in some areas such as fermentation (the traditional technology for making *sake*), Japanese research is of primary importance.

Today, tracking Japanese progress in high technology areas is both a necessary and rewarding process. Japanese academic institutions are very active; consequently, their results are published in scientific and technical journals and are presented at numerous meetings where more than 20,000 technical papers are presented orally every year. However, due principally to the language barrier, the results of academic research in Japan are not well known overseas. Many in the United States and in Europe are thus surprised by the sudden appearance of Japanese high technology products. The products are admired and enjoyed, but some are astonished at how suddenly these products appear.

With the series *Japanese Technology Reviews,* we present state-of-the-art Japanese technology in five fields:

> Electronics
>
> Computers and Communications
>
> New Materials
>
> Manufacturing Engineering
>
> Biotechnology

Each tract deals with one topic within each of these five fields and reviews both the present status and future prospects of the technology, mainly as seen from the Japanese perspective. Each author is an outstanding scientist or engineer actively engaged in relevant research and development.

The editors are confident that this series will not only give a bright and deep insight into Japanese technology but will also be useful for developing new technology of our readers' own concern.

As editor in chief, I would like to sincerely thank the members of the editorial board and the authors for their contributions to this series.

TOSHIAKI IKOMA

Preface

Electronic tuning is widely used in today's televisions for the selection of VHF, UHF, and CATV channels. Compared to conventional mechanical tuners, electronic tuners are superior in terms of size, speed of tuning, reliability, and integration of additional functions such as channel memory and remote control.

The key components that make possible the realization of electronic tuners are the voltage-variable capacitance diodes (varactor diodes) for the actual tuning, the band-switch diodes for channel selection, and the dual-gate FETs for front-end RF amplification. Monolithic integrated circuits have found increasing usage in electronic tuners.

Noise figure and power gain have been the major concern for high-frequency devices. However, low distortion is as important for electronic tuners, and newly developed design principles for low distortion are described in this book. Both noise figure and distortion are important parameters, since TV tuners must operate in both weak and strong signal-reception conditions. Low-distortion characteristics have been of special concern because of the relatively inferior immunity to interfering signals in electronic tuners. The capability of being mass produced is also essential for devices aimed at consumer TVs and VCRs.

The development of high-frequency devices for electronic tuners remains an important topic, since the market for TVs and VCRs will continue to expand as new varieties of products are introduced, including portable VCRs, high-definition TV/VCR systems, video printers, and computer-assisted audiovisual systems.

The author hopes that readers will find the contents of this book informative and enlightening, in terms of both high-frequency device development and system circuit design.

Acknowledgments

I am most grateful to Prof. T. Ikoma for providing me with the opportunity to write this book and for giving advice in organizing its contents. I am also grateful to Profs. T. Sugano, S. Takaba, M. Takagi, K. Tada, and H. Sakaki for their advice. In the research and development of semiconductor devices, cooperation between people having different kinds of technical experience is important as the work involved spans various fields, including materials science, process development, device fabrication, evaluation, circuit design, and systems application. I am most fortunate to have been able to work with numerous prominent colleagues in pursuing the performance and fabrication technology of these devices.

I would like to express sincere gratitude to my colleagues at Sony for their generosity in allowing me to incorporate their findings into this book. The study of the voltage-variable capacitance diode is joint work with Hideo Kubota, Yasuo Hayashi, Hisayoshi Yamoto, Haruo Sone, Nobumichi Okazaki, and Teruaki Aoki. The development of the diode for CATV tuners with a high capacitance ratio was done by Haruo Sone. The research on the band-switch diode was done with Hisayoshi Yamoto. The investigation into the dual-gate MOSFET was done with Akio Kayanuma, Kaoru Suzuki, Yukio Tsuda, Shinichi Saiki, and Osamu Yoneyama. The dual-gate GaAs MESFET was pursued with Shinichi Tanaka, Junichiro Kobayashi, Hajime Ohke, Kaoru Suzuki, Osamu Yoneyama, Hidemi Takakuwa, Tsuneo Aoki, and Masayoshi Kanazawa. The monolithic integrated circuit was jointly developed with Shinobu Turumaru, Takashi Yoshikawa, Kenichiro Kumamoto, Kazuo Watanabe, Shigeo Matsumoto, Kaoru Suzuki, and Yukio Tsuda.

A number of figures have been taken from the literature. Thanks go to Dr. A. G. K. Reddi and Prof. T. Hara for the figures in chapter 2. The GaAs diode described in this book was done by Dr. T. Hara and his group and I am very grateful to Dr. Hara for his permission to reproduce some of the results.

This work has been supported by helpful guidance and assistance from many people at Sony and I would like to thank them for their contributions. Makoto Kikuchi, Yoshiyuki Kawana, Hajime Yagi, Jun Numata, and Naozo Watanabe provided guidance in all aspects of the work. Akiyasu Ishitani, Takashi Shimada, Yoshio Oka, Naozo Watanabe, Tadao Kuwabara, Isao Kajino, Koji Hirano, Goro Miyake, Hideki Minami, Takashi Seki, and Shigeo Matumoto offered generous assistance.

This work has also been supported by members of the board of directors of Sony, namely, Masahiko Morizono, Masahiro Takahashi, and Akinori Takasaki. I thank them for their great support and understanding.

Sincere thanks also to Kou Togashi, who took the trouble of proofreading the English manuscript.

CHAPTER 1

Introduction

Abstract

Low distortion as well as low-noise characteristics as specific requirements of semiconductor devices used for electronic tuners are described.

1.1. Evolution of Electronic Tuners

The idea to utilize the reverse-biased diode as a variable capacitance for frequency tuning existed from the early days of semiconductor development.[1] Electronic tuners with voltage-variable capacitance diodes were first widely introduced in Europe.

About 1974, electronic tuners began to be employed for the main models of television sets both in the United States and in Japan.[2] In the first models, voltages for tuning were defined by potentiometers. However, because of a number of advantages that electronic tuners had, mechanical tuners were rapidly taken over.

Behind such changes was the development of highly productive voltage-variable capacitance diodes, as described later.

The emergence of LSI technology made flexible tuning possible, such as tuning by voltage synthesizers and frequency synthesizers. This technology made it possible to add a number of functions to television sets, which in turn made electronic tuners indispensable.

As UHF bands became widely used, voltage-variable capacitance diodes were improved to meet the requirement for high Q; low-noise active devices such as dual-gate MOSFETs and dual-gate GaAs MESFETs were introduced.

As CATV became widely available, the number of channels increased dramatically, exceeding 100. Voltage-variable capacitance diodes with large capacitance were required and were developed.

As for packages, surface-mount models replaced first-generation packages about 1983, and the demand for tape-and-reel shipments increased. Even smaller packages have been developed recently.

High-frequency semiconductor devices have always been and will continue to be the key components for technological improvement and market penetration of the new system.[3]

1.2. Expectations of Electronic Tuners

The reasons why electronic tuners were desired can be summarized as follows:

1. *Demand for high reliability.* Mechanical contacts often cause reliability problems. It was desired to eliminate mechanical moving parts to increase reliability.
2. *Demand for complex functions.* As the number of channels increased, and as VCRs became widely used, complex functions such as channel selection by programming became necessary. Thus, electronic tuning became necessary.
3. *Demand for smaller size.* As portable usage increased, smaller size became important.

1.3. Requirements for High-Frequency Semiconductor Devices

In mechanical tuners, variable capacitors or sets of switchable wire inductances are utilized as frequency-tuning devices. Variable capacitors are composed of sets of rotating plates that work as air-gap capacitors. The sets of wire inductances are selected by mechanical switches.

In electronic tuners, semiconductor devices are used for frequency tuning. Voltage-tunable capacitance diodes replace variable capacitors. Switches are replaced by band-switch diodes, which change the inductance value by acting as electronically controllable switches.

These semiconductor devices inherently have considerably high parasitic series resistance due to relatively low carrier concentration and limited carrier mobility. The series resistance contributes

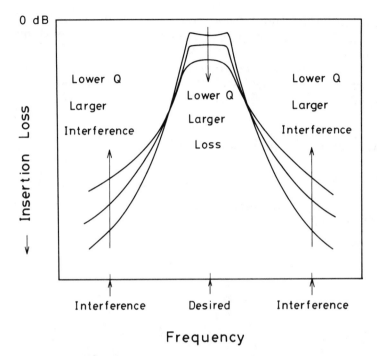

Figure 1. Band-pass characteristics of tuning circuits having different Q values. The lower the Q value, the larger the insertion loss and the larger the interfering signal passing the tuning circuit. The insertion loss gives rise to noise figure and the interfering signal causes cross-modulation distortion at the following amplifier or mixer.

to lowering the Q value of the resonant circuit, increasing the insertion loss of the circuit.

The situation is shown in Figure 1. As the Q value of a resonant circuit decreases, the insertion loss increases, and the sharpness of the cutoff characteristic degrades, allowing a larger level of undesired signals to pass through to the following stage.

For these reasons, active devices with lower noise figures and lower cross-modulation distortion are required in electronic tuners. Dual-gate MOSFETs and dual-gate GaAs MESFETs are widely employed for the first-stage amplifier of tuners because of their good gain-control characteristics. The improvement of cross-modulation distortion in the gain-reduction range is especially required.

Another origin of distortion lies in the voltage-variable capacitance diode itself. Diodes with low distortion are a fundamental requirement.

1.4. Examples of Tuner Circuits

In Figure 2, a UHF tuner with mechanical variable capacitors is shown. In Figure 3, a circuit diagram of a VHF tuner that employs a VHF tuner with mechanically switchable wire inductors is

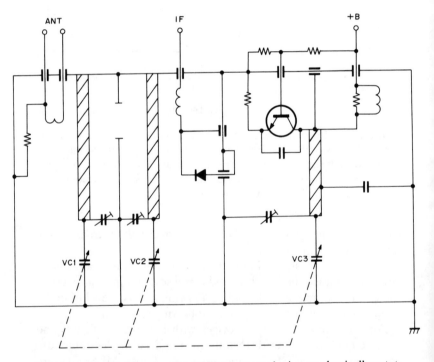

Figure 2. Circuit diagram of a UHF tuner employing mechanically rotating variable capacitors.

shown. The VHF tuner acts as an IF amplifier when a UHF signal is received to c or ipensate for the relatively low gain of the UHF tuner. The large number of mechanical switches was often the bottleneck to improving reliability.

In Figure 4, an example of an electronic tuner that contains both VHF and UHF parts is shown. A dual-gate MOSFET is used for the first-stage amplifier of the UHF tuner to compensate for the degradation of the noise figure due to the low Q value of the variable-capacitance diodes. Band-switch diodes are employed to switch the inductors in the tuner to cover the lower and upper VHF bands effectively. A monolithic IC acts as the VHF mixer and oscillator, and it also provides a matched impedance output to the IF stage. This IC acts as the IF amplifier when UHF channels are received.

The whole circuit is realized in a printed-circuit board that is then housed in a metal shield box. Mechanical switches and moving parts are eliminated.

References

1. M. H. Norwood and E. Shatz. Voltage Variable Capacitor Tuning: A Review. *Proc. IEEE* **56**, 5(1968): 788–798.
2. Y. Fujioka and K. Kumamoto. *Denshigijutu* **19**, 7(1975): 38–40.
3. S. Watanabe. Ph.D. dissertation, University of Tokyo, 1989.

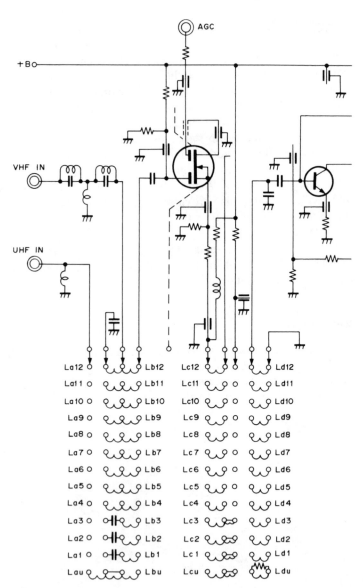

Figure 3. Circuit diagram of a VHF tuner employing mechanically switchable wire inductors for tuning.

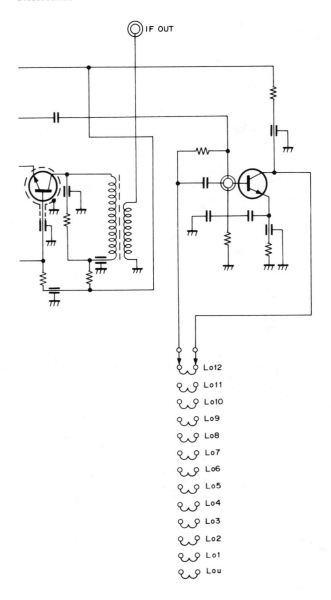

IF OUT

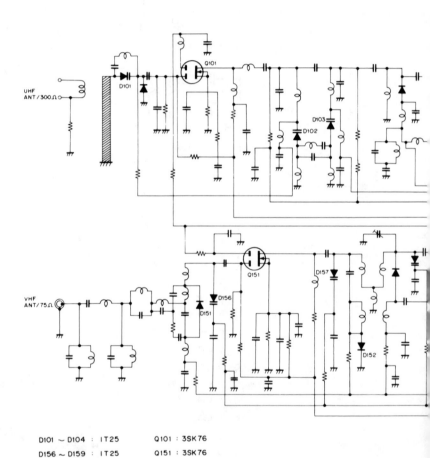

D101 ~ D104 : IT 25   Q101 : 3SK 76
D156 ~ D159 : IT 25   Q151 : 3SK 76
D151 ~ D155 : IT 26
                            IC-151 : CX-099

Figure 4. Circuit diagram of an electronic tuner containing both VHF and UHF parts.

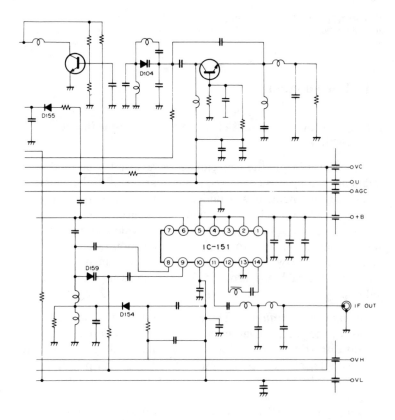

CHAPTER 2

Voltage-Variable Capacitance Diodes

Abstract

Various aspects of voltage-variable capacitance diodes, including analysis, design, and fabrication processes of devices with low distortion and high Q values, are discussed.

2.1. Introduction

2.1.1. Requirements of Voltage-Variable Capacitance Diodes

The idea of utilizing the capacitance of reverse-biased diodes for frequency tuning was conceived many years ago.[1] In the course of theory development, it was understood that a hyperabrupt junction is effective for achieving a high capacitance ratio.

However, the device did not come into wide use in consumer applications mainly due to immature production technology. The requirements for wide application of this device involved quite a few technological breakthroughs. The problems are summarized as follows:

1. *Nonlinear distortion.*[2–9] The capacitance of the diodes changes as the incoming signal modulates the bias, resulting in nonlinear distortion. The desirable capacitance–voltage (CV) curve was determined experimentally and the design principle was not clearly established. The design of CV curves that result in low nonlinear distortion was required.
2. *Low Q value due to parasitic series resistances.* The undepleted region under low reverse-bias conditions contributes to the rise in series resistance. There are a number of other resistive elements that also contribute to the series resistance. These series resistances degrade the Q value considerably compared with

that of mechanical variable capacitors. Therefore, these series resistances should be made small by process technologies to levels where the residual difference can be compensated for by other active devices.

3. *Nonuniformity of characteristics.* The resonant circuits in a tuner, such as those for the input stage, the output of the first-stage amplifier and the input of the mixer, and for the oscillator, should be tuned to the desired frequency at all the biasing voltages, or the tracking error should be within the required amount.

The primary cause of the tracking error is due to the different CV characteristics of the diodes. During measurement and sorting, the diodes are sorted into matched groups according to the CV curves. In reality, the number of measuring points is limited, and the number of sorted groups is also limited. Therefore, the uniformity of the CV curves of the devices is extremely important in view of productivity of the device. The nonuniformity of the epitaxial layer thickness also gives rise to high series resistance.

2.1.2. Basic Structure and Characteristics

The schematic of a voltage-variable capacitance diode is shown in Figure 5. An N^+ region with relatively high impurity concentration is formed for the hyperabrupt junction. The P^+ region is doped sufficiently high so that the depletion layer extends only into the N region.

Examples of the impurity-concentration profile of the hyperabrupt junction are shown in Figure 6. In the profile of Figure 6(b), the concentration changes in a stepwise manner, and the points of change (x_1 and x_2) correspond to the depths that determine the maximum and minimum capacitances, respectively, required for the coverage of the frequency range. However, this profile gives a sharp slope to the CV curve of the diode, resulting in high nonlinear distortion.[9]

In order to reduce the nonlinear distortion, a graded profile, as shown in Figure 6(a), is employed. How to obtain the impurity profile, which gives a CV curve with low distortion, is the subject of the following sections.

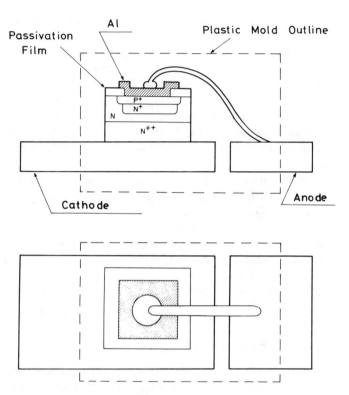

Figure 5. Schematic of a voltage-variable capacitance diode.

2.1.3. Capacitance Change and Frequency Coverage

Important characteristics of voltage-variable capacitance diodes are the maximum capacitance and the minimum capacitance, the series resistance, and the measure of nonlinear distortion.

The bias voltages that correspond to the maximum and minimum usable capacitances are usually specified as 2 volts and 25 volts, respectively, and the corresponding capacitances are designated as C_2 and C_{25}, respectively. Hence, the frequency ratio (f_{max}/f_{min}) is expressed as

$$f_{max}/f_{min} = (C_2/C_{25})^{1/2} \tag{2.1}$$

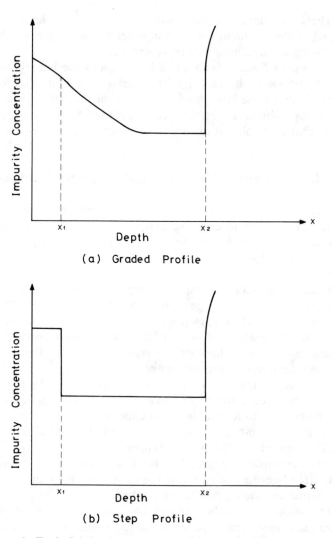

Figure 6. Typical impurity-concentration profiles of the hyperabrupt junction.

The lower limit is determined by the amplitude limit of the signal or the oscillation voltage in order to keep the diode from becoming forward biased by these voltages.

The upper limit is determined by design of the biasing circuit and the diode itself. The high breakdown voltage, needed to assure a high reverse bias, poses a more difficult reliability problem. If the upper limit is low, a sharp change in the CV curve is required and again the nonlinear distortion increases.

2.1.4. Uniformity of CV Curves and Frequency Tracking

In order to assure good matching of different diodes used in a tuner, the capacitances at several bias voltages are measured and the diodes sorted into different groups. Within a group, the relative spread of capacitance is guaranteed to be within 2% to 2.5%.

If the capacitance distribution is large, the measuring efficiency becomes low, and a large number of devices can be left unmatched. One major reason why diode tuning has not become dominant is the poor uniformity of the device characteristics due to immature process technology, preventing the realization of low-cost devices required for wide usage.

With the advent of ion-implantation technology, the achievement of required uniformity came within reach. The application of this technology to form a desirable impurity-concentration profile became the primary goal. An early work by R. A. Moline and G. W. Reutlinger reported the use of channeling,[10] which was so sensitive to the implanting direction that it was supposed to be unusable. In a later section, a technique to improve the uniformity of the partial-channeling profile is described.

The thickness and the impurity concentration of the epitaxial layer are also subject to variation, adding to the nonuniformity of the CV curve. The variation of the thickness also contributes heavily to the variation of the series resistance, increasing the maximum guaranteeable resistance and the average series resistance.

2.1.5. Considerations for Reduction of the Series Resistance

Another important reason why diode tuning did not become popular was the low Q value of the diode capacitor compared with that of air-gap variable capacitors. In order to reduce the series resistance of the N-type layer between the P^+ and the substrate N^{+2} layers, the impurity concentration should be increased to the extent allowed by the reverse breakdown voltage.

The larger the capacitance change required, the larger the series resistance of the N-type layer, for the impurity concentration should be accordingly decreased and the thickness increased. Therefore, any parasitic capacitance, such as that of the package or the junction, decreases the overall capacitance ratio and must be kept to a minimum, thereby minimizing the required series resistance as well.

In addition, there are a number of parasitic resistances specific to the diode. The physical interpretation of resistive elements must be analyzed in order to obtain a diode design with low series resistance.

2.2. Nonlinear Distortion of Voltage-Variable Capacitance Diodes

2.2.1. The Effects of Nonlinear Distortion

It has been reported that nonlinear distortion of voltage-variable capacitors causes various undesirable effects, such as a shift in the resonance frequency,[4] cross modulation,[5] and intermodulation.[2-9,11]

In this section, the nonlinear effects of a voltage-variable capacitor in a series resonant circuit are analyzed. It is shown that by appropriate approximation, the measure of nonlinear distortion is expressed in terms of the CV curve. The measure of nonlinear distortion corresponds to the nonlinear effects caused by the capacitor.[2,3,5,6,9]

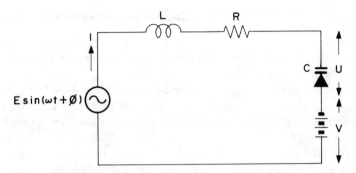

Figure 7. Series resonant circuit containing a voltage-variable capacitor.

It is also shown that one can derive a CV curve that gives a zero value of the measure of nonlinear distortion and that the curve is expressed by a simple analytical function. An impurity profile that gives the previously stated CV curve can also be derived, indicating that such a profile is a good design target of a voltage-variable capacitance diode.[3,5,6,9]

2.2.2. Analysis[3,5,6]

In Figure 7, a series resonant circuit containing a voltage-variable capacitor is shown. The small signal voltage that appears at the terminals of the capacitor is designated as U and the inductance as L.

If the bias voltage were designated as V, then the capacitance can be expressed by the following equation in the form of a Taylor series expansion:

$$C(V + U) = C(V) + \frac{dC}{dV} U + \frac{1}{2}\frac{d^2C}{dV^2} U^2 + \cdots$$
$$= C_0(1 + a_1 U + a_2 U^2 + \cdots) \tag{2-2}$$

where $C_0 = C(V)$

$$a_1 = \frac{1}{C_0}\frac{dC}{dV}$$

$$a_2 = \frac{1}{2C_0}\frac{d^2C}{dV^2}$$

In this analysis, it is assumed that the amplitude of the signal voltage is not very large, and that the nonlinearity is sufficiently small. Then,

$$| a_1 U | \gg | a_2 U^2 | \gg \cdots \qquad (2\text{-}3)$$

The differential equation for the circuit is given as

$$L \frac{dL}{dt} + RI \int \frac{I}{C} \, dt = E \sin (\omega t + \phi) \qquad (2\text{-}4)$$

Equation (2-4) is transformed into the following:

$$LC \frac{d^2 U}{dt^2} + \left(RC + L \frac{dC}{dt} \right) \frac{dU}{dt} + U = E \sin (\omega t + \phi) \qquad (2\text{-}5)$$

Under the assumption that the signal amplitude and the nonlinearity are both small, the amplitude of the harmonics decreases rapidly as the degree of the harmonics becomes higher. Therefore, only the second order $(2\omega t)$ need be considered.

If U is expressed as in Equation (2-6) and substituted into Equation (2-2), then Equation (2-7) is obtained for C.

$$U = U_1 \sin \omega t + e_2 \cos 2\omega t + u_2 \sin 2\omega t + \cdots \qquad (2\text{-}6)$$

$$C = C_0 \left[1 + \frac{a_2}{2} (U_1^2 + e_2^2 + u_2^2) + a_2 U_1 u_2 \cos \omega t + (a_1 U_1 - a_2 U_1 e_2) \right.$$

$$\left. \times \sin \omega t + \left(a_1 e_2 - \frac{a_2 U_1^2}{2} \right) \cos 2\omega t + a_1 u_2 \sin 2\omega t + \cdots \right] \qquad (2\text{-}7)$$

Substituting Equations (2-6) and (2-7) into Equation (2-3), and comparing the coefficients of $\cos 2\omega t$ and $\sin 2\omega t$, we obtain the following equations:

$$e_2 = \frac{a_1}{3} U_1^2 \qquad (2\text{-}8)$$

$$| u_2 | \ll | e_2 | \qquad (2\text{-}9)$$

In deriving Equations (2-8) and (2-9), the following approximations are made:

$$\omega_0^2 L C_0 = 1$$

$$Q = \frac{\omega_0 L}{R} \doteq \frac{\omega L}{R} \gg 1$$

$$\frac{\omega - \omega_0}{\omega_0} \ll 1$$

The equation for the fundamental frequency terms is

$$U_1 \left\{ 1 - \left(\frac{\omega}{\omega_0} \right)^2 \left[1 + \left(\frac{a_2}{4} - \frac{a_1^2}{6} \right) U_1^2 \right] \right\} \sin \omega t + \frac{U_1}{Q} \cos \omega t$$
$$= E \sin (\omega t + \phi)$$

$$(2\text{-}10)$$

The equation for constant capacitance is

$$U_1 \left[1 - \left(\frac{\omega}{\omega_0} \right)^2 \right] \sin \omega t + \frac{U_1}{Q} \cos \omega t = E \sin (\omega t + \phi) \qquad (2\text{-}11)$$

Equation (2-11) is equivalent to Equation (2-10) if the constant capacitance C_0 is changed to C:

$$C = C_0 (1 + \xi U_1^2) \qquad (2\text{-}12)$$

where

$$\xi = \frac{a_2}{4} - \frac{a_1^2}{6}$$

$$= \frac{1}{8} \left[\frac{d^2 C}{dV^2} - \frac{4}{3} \left(\frac{dC}{dV} \right)^2 \right] \qquad (2\text{-}13)$$

Therefore, the resonance characteristics of the circuit shown in Figure 7 can be regarded as those with the capacitance increased by the factor of ξU_1^2. Therefore, the shift of the resonant frequency is

$$\frac{\omega_1 - \omega_0}{\omega_0} = \frac{\xi}{2} U_1^2$$

$$= \frac{1}{16} \left[\frac{d^2 C}{dV^2} - \frac{4}{3} \left(\frac{dC}{dV} \right)^2 \right] U_1^2 \qquad (2\text{-}14)$$

The effect of nonlinear distortion is thus proportional to the measure of distortion (ξ) and the square of the amplitude of the applied signal (U_1^2).[3,5,6]

The equation for the amplitude of the fundamental signal can be expressed as

$$U_1^2 \left[1 - \left(\frac{\omega}{\omega_0} \right)^2 (1 + \xi U_1^2) \right]^2 + \left(\frac{U_1}{Q} \right)^2 = E^2 \tag{2-15}$$

2.2.3. The Measure of Nonlinear Distortion

In the case of a parallel resonance circuit, as shown in Figure 8, the differential equation is derived as

$$LC \frac{d^2 U}{dt} + \left(\frac{L}{R} + L \frac{dC}{dt} \right) \frac{dU}{dt} + U = -\omega LI \sin (\omega t + \phi) \tag{2-16}$$

The form of the equation is basically the same as Equation (2-5) for the case of a series resonant circuit. A similar calculation can be performed yielding the equivalent nonlinear effect, as shown in Equations (2-12) and (2-13) with ξ as the measure of nonlinear distortion.[2]

For a parallel resonant circuit, a calculation including harmonic terms as high as 3ω is reported by А. М. МАЛЛЯР.[11] He expressed U and C as shown in Equations (2-17) and (2-18), and obtained ξ, which corresponds to the amount of the resonance frequency shift, as shown in Equation (2-19).

$$U = U_1 \cos \omega t + u_2 \cos 2\omega t + e_2 \sin 2\omega t \tag{2-17}$$

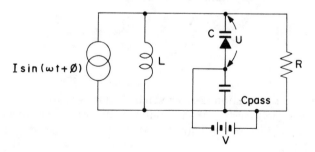

Figure 8. Parallel resonant circuit containing a voltage-variable capacitor.

$$C = C_0(b_0 + b_1 \cos \omega t + b_2 \cos 2\omega t$$
$$+ b_3 \cos 3\omega t + \cdots) \tag{2-18}$$

$$\xi = b_0 - 1 - \frac{b_2}{2} - \frac{(b_1 - b_3)^2}{3 + 4(b_0 - 1)} \tag{2-19}$$

By comparing Equations (2-5) and (2-18), and neglecting relatively small terms, the following equations are obtained for b_0, b_1, and b_2 in terms of a_1 and a_2 in Equation (2-5):

$$b_0 = 1 + \frac{a_2}{2} U_1^2 \tag{2-20}$$

$$b_1 = a_1 U_1 \tag{2-21}$$

$$b_2 = \frac{a_2}{2} U_1^2 - \frac{a_1^2}{3} U_1^2 \tag{2-22}$$

Assuming that b_3 is sufficiently smaller than b_1, and substituting Equations (2-20) to (2-22) into Equation (2-19), we obtain Equation (2-13). Therefore, Equation (2-13) is an approximate form of Equation (2-19).

The advantages of Equation (2-13) are that the measure of nonlinear distortion (ξ) can be easily calculated from the CV curve and that the differential equation obtained by setting ξ to zero can be solved analytically, yielding an effectively distortionless curve.

Y. Ninomiya studied the effects of nonlinear distortion of voltage-variable capacitors. He gave N in Equation (2-23) as the factor of nonlinear distortion,[2] and showed that the effects of nonlinear distortion such as the resonant-frequency shift and cross modulation can be expressed simply by the value of ξ.

$$N = Qp(QE)^2 \tag{2-23}$$

where

$$p = 3a_2 \tag{2-24}$$

Factor p in Equation (2-23) given by Ninomiya corresponds to 4ξ and gives the measure of the nonlinear distortion. However, the expression given in Equation (2-24) should be replaced by Equation (2-13).

R. G. Meyer and M. L. Stephens expressed the CV curve as

shown in Equation (2-25) and calculated the third-order intermodulation as given in Equation (2-26).[7]

$$C(V) = \frac{K}{(\phi_D + V)^n} \tag{2-25}$$

$$IM_3 = \frac{Q^2 a_2 P}{2\omega_0 C_0}\left(1 - \frac{2n}{1.5(n + 1)}\right) \tag{2-26}$$

where

$$P = \frac{U_1^2}{2R}$$

Measured data are also given and show good correspondence. Equation (2-26) can be expressed in terms of a_1 and a_2 in Equation (2-2), yielding Equation (2-27) for the third-order intermodulation.

$$IM_3 = QU_1^2\left(\frac{a_2}{4} - \frac{a_1^2}{6}\right)$$
$$= \xi QU_1^2 \tag{2-27}$$

Therefore, ξ can be regarded as the measure of the third-order intermodulation. The advantages of the expression given in Equation (2-13) are the same as previously described.

2.2.4. Distortionless Voltage-Variable Capacitance Diodes

A design of voltage-variable capacitance diodes that gives ξ a zero value is given in Equation (2-13). Setting ξ to zero, we obtain the following differential equation:

$$\frac{d^2 C}{dV^2} = \frac{4}{3}\left(\frac{dC}{dV}\right)^2 \tag{2-28}$$

Equation (2-28) can be solved analytically, giving Equation (2-29):

$$C = A(V + \Phi)^{-3} \tag{2-29}$$

where A and Φ are integration constants.

In order to obtain a CV curve commonly used for TV tuners,

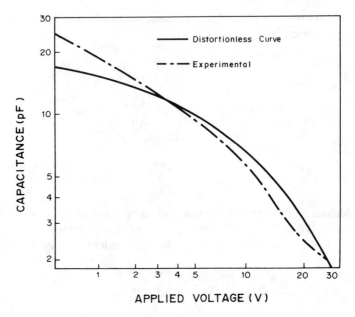

Figure 9. Calculated distortionless CV curve and a measured curve of 1T18.

capacitances at biases of 2 and 25 volts are given as 15 and 2.25 pF, respectively. Then the following values are calculated for both A and Φ.

$$A = 2.66 \times 10^5 (\text{pF} \cdot \text{V}^{-3})$$
$$\Phi = 24.1 \text{ volts}$$

The calculated curve is shown in Figure 9, where that of 1T18, commercially available from Sony, is also plotted for comparison.

The impurity-concentration profile of a hyperabrupt junction is shown in Figure 10, where the distance from the junction is placed at the x axis and that of the edge of the depletion layer at x_m. If the area is designated as S and the dielectric constant as ε, the capacitance C can be expressed as

$$C = \frac{\varepsilon S}{x_m} \tag{2-30}$$

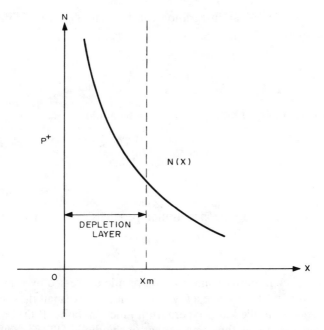

Figure 10. Schematic impurity-concentration profile of a P$^+$N hyperabrupt junction diode. The concentration of the P$^+$ layer is sufficiently large so that the spread of the depletion layer into the N layer need only be considered.

Then, substituting this into Equation (2-30), one obtains

$$V + \Phi = \left(\frac{A}{ES}\right)^{1/3} x_m^{1/3} \tag{2-31}$$

The applied voltage and the impurity concentration are related as follows:

$$V = \frac{q}{\varepsilon} \int_0^{x_m} \int_x^{x_m} N(x)\, dx\, dx' - \phi_D \tag{2-32}$$

If the impurity concentration is expressed as Equation (2-23) and is substituted into Equation (2-32), Equation (2-34) is obtained with B as the integration constant and ϕ_D as the diffusion potential.

$$N(x) = N_0 x^{-n} \tag{2-33}$$

$$V = \frac{q}{\varepsilon} \cdot \frac{N_0 x_m^{-n+2}}{n-2} + B - \phi_D \tag{2-34}$$

For Equations (2-28) and (2-31) to be equal,

$$N_0 = \frac{\varepsilon}{3q} \left(\frac{A}{\varepsilon S}\right)^{1/3} \tag{2-35}$$

$$n = -\frac{5}{3} \tag{2-36}$$

Then, the impurity-concentration profile is expressed as

$$N(x) = \frac{\varepsilon}{3q} \left(\frac{A}{\varepsilon S}\right)^{1/3} x^{-5/3} \tag{2-37}$$

Thus, the impurity-concentration profile expressed by Equation (2-37) is expected to give a CV curve with a very small third-order distortion if applied to a hyperabrupt junction diode. If the area of the junction (S) is the same as that of the diode (1T18) shown in Figure 9 ($S = 3.9 \times 10^{-4}$ cm^2), then the profile is calculated as

$$N(x) = 4.16 \times 10^9 x_m^{-5/3} \tag{2-38}$$

Equation (2-37) is plotted in Figure 11 with the impurity profile of the reference diode (1T18).

The impurity concentration given in Equation (2-38) becomes infinitely large as x approaches zero in the junction. In Equation (2-28), applied voltage V could result in a large negative value. However, in practice, V should be zero at its minimum. Therefore, from Equation (2-28), the maximum capacitance that corresponds to an applied voltage of the zero value is determined. The maximum capacitance then gives the minimum value of the depth of the depletion-layer edge through Equation (2-30).

This means that without an applied bias, the depletion layer should spread to the depth corresponding to the maximum capacitance. Therefore, the impurity concentration should be zero. At least the region should be intrinsic in practice. There are a few

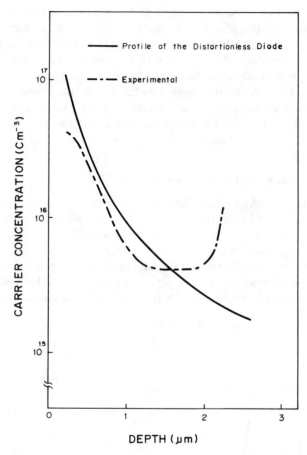

Figure 11. Calculated impurity profile of a distortionless diode and that of a reference diode (1T18).

practical means to realize very low or effectively zero impurity concentration near the junction, resulting in very low distortion around low applied voltages.

However, there have been no reports on hyperabrupt junction diodes that have a very low impurity concentration near the junction, and the realization must wait for the results of future work.

If the value of n in Equation (2-26) is 3, the intermodulation distortion becomes zero. This corresponds to the distortionless curve given in Equation (2-29), where the power index is -3. However, Φ is introduced as an integration constant in Equation (2-29), and ϕ_D is defined as the diffusion potential in Equation (2-26).

Therefore, the solution based on the differential equation given by Equation (2-13) has the advantage of giving a more general CV curve that is free of third-order distortion, and also in defining an impurity-concentration profile.

2.2.5. Experiment on the Frequency Shift[3]

The measure of nonlinear distortion (ξ) expressed in Equation (2-13) is calculated from the CV curve for the diode (1T18) shown in Figures 10 and 11 and is plotted in Figure 12. In Figure 12, the values of ξ calculated from the experimental data are also shown.

The experiment for measuring the resonant-frequency shift due to the nonlinear effect was performed with the series resonant circuit shown in Figure 13. The frequency was chosen as 100 MHz

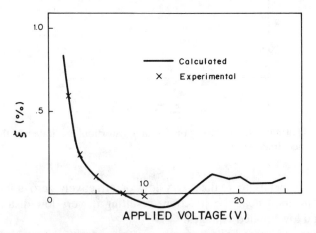

Figure 12. Measure of nonlinear distortion calculated from the CV curve of the reference diode (1T18). The values of ξ calculated from the measured frequency shift are also shown.

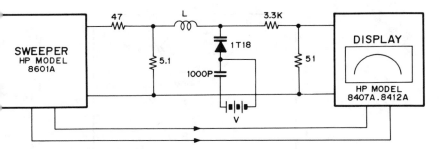

Figure 13. Circuit configuration for measuring the resonant-frequency shift due to the nonlinear distortion effect.

and both the signal voltage (U_1), which appears at the terminals of the diode, and the resonant-frequency shift (Δf) were measured. The bias voltages were 2.0, 3.5, 5, and 7.8 volts.

The results are shown in Figure 14, with $\Delta f/f$ along the Y axis and U_1 along the X. The values of ξ calculated from these results through Equation (2-13) are plotted in Figure 12. The values of ξ calculated from the original CV curve of the diode are also shown in Figure 14.

We can see that the shifts of the resonant frequency are proportional to the square of the signal amplitude (U_1^2), and that the experimental data fit the theoretically calculated values very well. Therefore, it can be expected that the CV curve and, consequently, the impurity-concentration profile, which gives small values of ξ, will be a practical solution for a voltage-variable capacitance diode with small nonlinear distortion effects.

2.3. ³¹P⁺² Ion Implantation for Low-Distortion Diodes[12]

2.3.1. Ion Implantation as a Fabrication Technology

In the practical fabrication of voltage-variable capacitance diodes, the uniformity of characteristics is not only important within a wafer, but also among different wafers and different lots, because

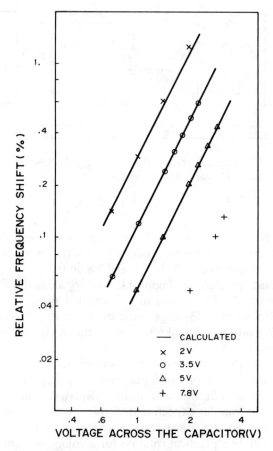

Figure 14. Measured resonant-frequency shifts at various reverse-bias voltages as a function of the applied signal voltage.

they are used in different resonance circuits in tuners as matched sets of diodes.

Ion implantation is a promising technology for those devices that require uniformity of impurity-concentration profiles. However, the implanted profile is basically gaussian and causes nonlinear distortion in the transition region from the implanted profile to the epitaxial layer. Channeling was employed by R. A. Moline and G. F. Foxall for fabrication of a hyperabrupt junction voltage-variable capacitance diode,[10] but the channeling profile is so sensitive to the implantation angle that it is not suitable for practical application. In this section, we review "partial channeling," which occurs by implantation from a direction tilted slightly to a specific direction from the low-index crystallographic axis such as ⟨111⟩.

In partial channeling, the profile is slightly affected by channeling and the impurity can be introduced into a deeper region. It will be shown that in an implantation from the direction of partial channeling, the profile can be made uniform by implantation through a thin layer of silicon dioxide (SiO_2).

It will also be shown that partial channeling is effective in fabrication of low-distortion voltage-variable capacitance diodes.

2.3.2. Partial Channeling and Uniformity

It was reported by V. G. K. Reddi and J. D. Sansbury that the dechanneling effect depends on the direction in which the beam is tilted and that the partial-channeling profile is obtained by tilting by appropriate angles into specific directions (for example, to the [11$\bar{2}$] direction by 5 to 10 degrees) in ion implantation into the (111) silicon crystal. Figure 15 shows the reported data.[13]

The partial channeling occurs in the direction between the [111] axis channeling and the adjacent low-index axis channeling, and the dependence of the profile on the direction is much smaller than the normal axis channeling. Partial channeling is explained by the existence of closely adjacent axis channeling and the existence of plane channeling.

2.3.3. Experimental Methods

$^{31}P^{+2}$ was employed as the ion and was implanted into an N-type (111) silicon crystal with a resistivity between 60 and 80 ohm-cm after acceleration by an energy of 400 keV equivalent. Annealing was performed at a temperature of 900°C in an nitrogen atmosphere for 10 minutes. Schottky diodes were then fabricated by evaporating gold. The carrier concentration was calculated from the measured CV curve.

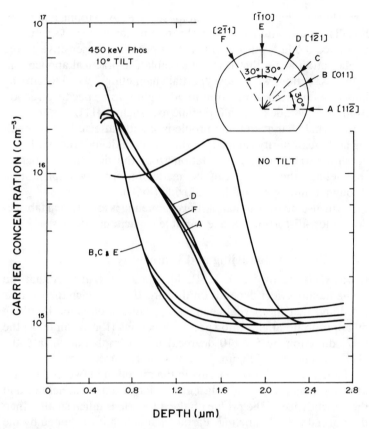

Figure 15. Carrier-concentration profiles resulting from 450-keV $^{31}P^{+2}$ implantation into silicon tilted by 10 degrees away from the [111] axis in various directions. The nominal dose was 1.5×10^{12} atoms cm^{-2}. (After V. G. K. Reddi and J. D. Sansbury.[13])

Ion implantation was performed from two directions and the profiles were compared. One is the direction in which the partial channeling is known to occur, as described by V. G. K. Reddi et al.,[13] that is, 6 to 7 degrees away from the [111] to [11$\bar{2}$] direction. The other is 6 to 7 degrees away from the [111] to [$\bar{1}$10] direction, which is known to result in a random profile or in a gaussian profile. The ion implantation was done through the thin silicon-dioxide film on silicon wafers with a thickness of 200 Å. The SiO_2 film was fabricated by oxidation in dry oxygen at 900°C.

The ion beams are usually scanned by electrostatic deflection. In this case, the implantation angle changes by several degrees within a silicon wafer. In the implantation by Accelerators Model 300A, which was employed in all the following experiments, the difference of the implantation angle between the center and the edge of a silicon wafer of 50 mmφ is about 1.5 degrees.

In order to investigate the effect of SiO_2 films on silicon wafers on the uniformity of the profile, ion implantation was performed from the partial-channeling direction through four different thicknesses of SiO_2, ranging from zero to 560Å. The carrier-concentration profile was measured at five points in a wafer with a diameter of 50 mm; one point is at the center and the other four are on the circle with a radius of 18 mm from the center of the wafer, placed at each end of two diameters of the circle perpendicular and parallel to the facet.

Voltage-variable capacitance diodes were fabricated in order to compare the characteristics of the diodes made by the partial-channeling profile and the random profile.

The fabrication process was as follows and is shown in Figure 16.

1. $^{31}P^{+2}$ was implanted through SiO_2 (200 Å) at an energy of 400 keV and with a dose of 6×10^{12} cm^{-2} into silicon wafers with N-type epitaxial layers ($\rho \simeq 1.0$ ohm-cm) on N-type low-resistivity substrates.

2. The surface of the wafer was covered with SiO_2 to prevent out-diffusion at phosphorus. The annealing and diffusion was performed at 1100°C for 70 minutes in a nitrogen atmosphere.

3. The high concentration P+ layer was formed by thermal diffusion followed by evaporation of aluminum for surface metallization and gold for the reverse side.

2.3.4. The Results

2.3.4.1. Dependence of Carrier-Concentration Profiles on Direction

In Figure 17, the carrier-concentration profiles due to implantation from the direction of the partial channeling and that of random

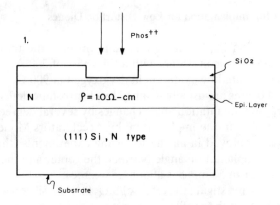

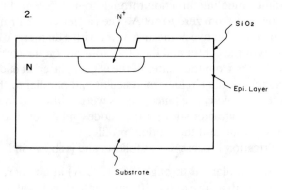

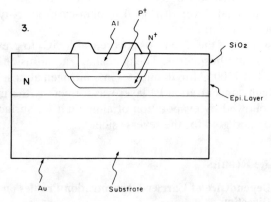

Figure 16. Fabrication process of the voltage-variable capacitance diode using $^{31}P^{+2}$ implantation.

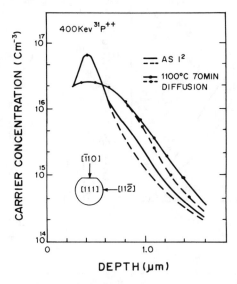

Figure 17. Carrier-concentration profiles resulting from 400-keV $^{31}P^{+2}$ implantation into silicon tilted by 6 to 7 degrees away from the [111] axis and after annealing at 1100°C. Solid curves correspond to tilting about the [$\bar{1}$10] axis and the dashed curves correspond to tilting about the [11$\bar{2}$] axis. The nominal dose was 1.7×10^{12} atoms cm^{-2}.

profiles are shown. The profiles after diffusion for 70 minutes at 1100°C are also shown.

It can be seen that with partial channeling, the impurity is introduced into deeper regions than with random profiles. Partial channeling is an effective method for the fabrication of voltage-variable capacitance diodes where the introduction of impurities into deep regions by limited, finite acceleration energies is required and a smooth transition of profiles between that of ion implantation and that of the epitaxial layer must be maintained.

2.3.4.2. Effect of SiO$_2$ Thickness on the Uniformity of Profile

In Figure 18, the deviations of the carrier-concentration profiles obtained by $^{31}P^{+2}$ implantations through SiO$_2$ with thicknesses of 0, 100, 200, and 560 Å, respectively, are shown. Ions were implanted

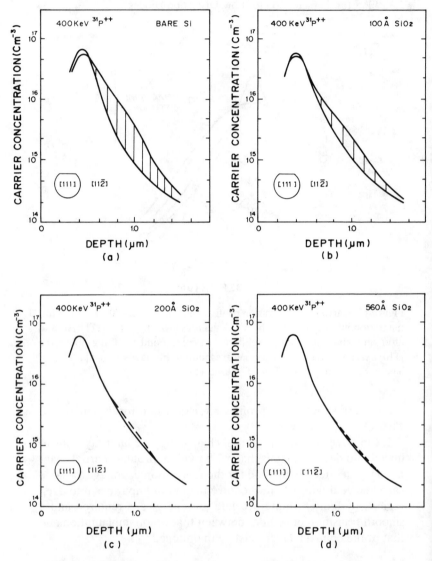

Figure 18. Uniformity of the carrier-concentration profile in a wafer resulting from 400-keV $^{31}P^{+2}$ implantation into silicon tilted about the $[\bar{1}10]$ axis by 6 to 7 degrees away from the $[111]$ axis through SiO_2 layers of various thicknesses: (a) bare silicon, (b) 100 Å, (c) 200 Å and (d) 560 Å. The nominal dose was 1.7×10^{12} atoms cm^{-2}.

from the direction of partial channeling, that is, 6 to 7 degrees away from the [111] to [11$\bar{2}$] direction. As the thickness of the oxide increases, the profile by partial channeling becomes more uniform within the wafer, and with an oxide thickness of 560 Å, the deviation becomes very small at the expense of the partial-channeling effect itself.

2.3.3.3. Characteristics of Fabricated Diodes

The CV curves of the two diodes made by ³¹P⁺² implantation from the direction of partial channeling and from that yielding the random profile are shown in Figure 19. The fabrication process was described in Figure 16. The distortionless curve calculated in the

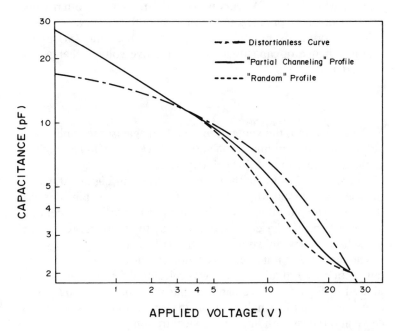

Figure 19. CV curves resulting from ³¹P⁺² implantation from the direction of partial channeling and from that yielding the random profile. The distortionless curve is plotted for reference.

previous section is also plotted, revealing that the CV curve obtained by partial channeling is closer to the distortionless curve.

The direction for the random profile was 6 to 7 degrees away from the [111] to [$\bar{1}$10] direction and that for the partial-channeling profile was 6 to 7 degrees away from the [111] to [11$\bar{2}$] direction.

As for the evaluation of the uniformity of the capacitance, it is difficult to separate the influence of nonuniformity of the epitaxial layer. In the simple comparison of the capacitances at the bias voltage of 12 volts for the five points within a wafer for measurement defined previously, that of the fabricated diode by the random profile showed a deviation of ±12% and that by the partial-channeling profile showed ±7%, assuring sufficient uniformity for the latter.

It has been known empirically that voltage-variable capacitance diodes, which have CV curves with sharp changes, often cause tracking errors in practice and require smooth CV curves.

The smoothness has been evaluated by the ratio (m) of the relative change of capacitance to the relative voltage change, expressed as

$$m = -\frac{dC}{C}\frac{V}{dV}$$

(2-38)

For evaluation, the value of m is plotted against the applied bias voltage on a logarithmic graph and we check to see if m is confined to the area determined through experience.

The values of m for both diodes are shown in Figure 20, where the specified area in which the plot of m should be confined is also depicted. The plot of m of the device made by partial channeling is within the specified area, whereas that by the random profile does not satisfy the specification.

The measure of nonlinear distortion (ξ) introduced in the previous section by Equation (2-13) was calculated for both diodes and plotted in Figure 21. We can see that the diode made by partial channeling exhibits smaller values of ξ, reflecting the smooth CV curve and the smooth impurity-concentration profile.

R. A. Moline et al. reported on additional ion implantation utilizing the channeling along the [111] axis after making the main profile by the random profile. The channeling was introduced to improve the impurity profile around the transition region between

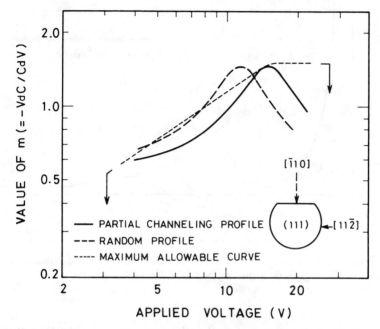

Figure 20. The m values of the voltage-variable capacitance diodes with partially-channeled and random profiles. 400-keV $^{31}P^{+2}$ was implanted through 200 Å thick SiO_2 from the directions 6 to 7 degrees away from the [111] about [11$\bar{0}$] axis for the "partially channeled" and 6 to 7 degrees away about [11$\bar{2}$] for the "random" profiles.

the implanted and the epitaxial layer, thus improving the CV characteristics.[10]

The method described here has the advantage of making the profile definition possible by a single implantation procedure.

2.4. Consideration of the Series Resistance

2.4.1. Series Resistance and the Q Value

The Q value of a voltage-variable capacitance diode is expressed as

$$Q = \frac{1}{\omega C R_s}$$

(2-39)

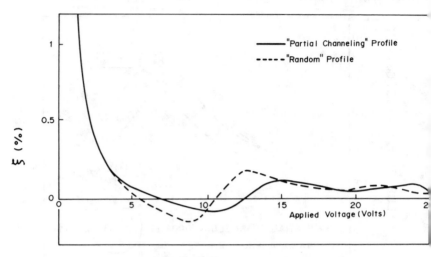

Figure 21. Measure of the nonlinearity of the voltage-variable capacitance diode with a partial-channeling profile and a random profile.

where C is the capacitance, R_s is the series resistance, and ω is the angular frequency.

When the capacitance is at its maximum value, the width of the undepleted region is at a maximum and the series resistance also is at its maximum value, hence giving the lowest Q value.

Reduction of the relatively large resistance compared with that of the mechanically variable capacitor has been the drawback for electronic tuners.

In this section, the various contributing factors to the series resistance are investigated for the 1T6 diode (Sony) in order to draw design principles for high-Q diodes.

In Figure 22, the cross section of the 1T6 diode is shown, indicating the various factors of series resistance. The diode has a mesa structure, minimizing the parasitic capacitance of the P⁺ layer spreading outside of the N⁺ layer in a planar structure. Table 1 shows the structural parameters and typical characteristics of the diode.

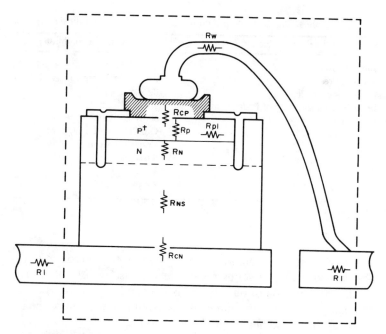

Figure 22. Cross section of the 1T6 diode and various contributing elements to the series resistance.

2.4.2. Estimation of Contributing Factors

2.4.2.1. The Epitaxial Layer with Flat Impurity Concentration

The carrier-concentration profile of the diode with relatively low series resistance is calculated from the CV curve and is shown in Figure 23. The whole impurity-concentration profile of the N layer of this diode is defined by the epitaxial growth, in order to obtain a smooth CV profile.

The contribution from the epitaxial layer with a relatively flat concentration profile is the largest and calculated to be about 0.30 ohm, corresponding to a layer of 1.2 microns with a resistivity of 1.0 ohm-cm.

Table 1. Parameters and characteristics of the 1T6 diode

Parameters	Value or condition
Junction area	4×10^{-4} cm^2 (200 μm $\times$ 200 μm)
Substrate resistivity	0.001–0.003 ohm-cm (As doped)
P$^+$ junction depth	0.5 μm
Impurity concentration of P$^+$ layer surface	1.8×10^{20} cm^{-3}
Bonding wire	50 μmϕ (Au)
Outer leads	Gold-plated (2 μm^t) covar
Package	Epoxy plastic mold
Series resistance at 470 MHz	0.6 ohms
Capacitance at 2 volts	15 pF
Capacitance at 25 volts	2.4 pF

A deviation of 0.1 ohm-cm in resistivity or 0.3 micron in thickness would lead to a series resistance increase of 0.12 ohm. The maximum deviation of both the resistivity and the thickness of the epitaxial layer then determines the maximum deviation of the series resistance, and hence the specification.

Therefore, if the deviation can be controlled to a small value, then the average series resistance due to the epitaxial layer with a flat concentration can be made small. A deviation of 0.12 ohm should be reduced to half, which is 0.06 ohm, by means of tight process control during fabrication.

2.4.2.2. The Hyperabrupt N$^+$ Layer

The contribution from the hyperabrupt N$^+$ layer is 0.07 ohm. To reduce the resistance, the concentration should be increased, resulting in a decrease of the breakdown voltage. The impurity profile should satisfy the requirement of low distortion. Therefore, not much can be done for the N$^+$ layer.

2.4.2.3. The Substrate with Low Resistivity (R_{NS})

The substrate is doped with arsenic to a very high concentration, yielding a resistivity between 0.001 to 0.003 ohm-cm. The thickness is 150 microns. The substrate is square, having 400 microns

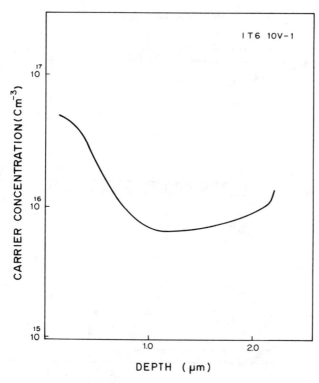

Figure 23. Carrier-concentration profile of the 1T6 diode.

per side, the bottom covered by evaporated gold. The upper junction area is square, 200 microns per side.

The contribution from the substrate can be roughly estimated by assuming that it can be divided into three square-shaped plates, as shown in Figure 24, and that the resistance is the sum of the resistances between the upper and bottom plates.

The resistance of the substrates with a thickness of 100 microns is also calculated for the cases of different substrate resistivities. The results are summarized in Table 2.

If the substrate is thinned to 100 microns in thickness and the resistivity is under 0.002 ohm-cm, then the contribution from the

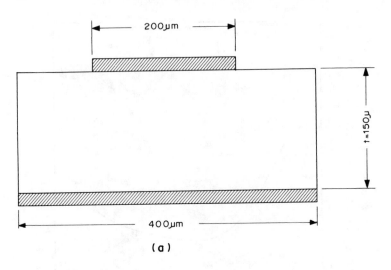

(a)

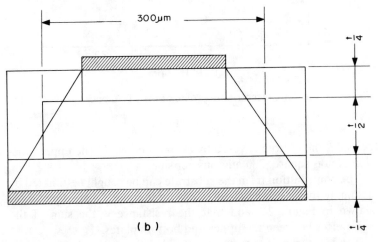

(b)

Figure 24. Approximation model for calculating the resistance of the substrate.

Table 2. Resistance of the substrate

ρ (ohm-cm)	0.001	0.002	0.003
R_{SN} (150 μm')	0.02	0.04	0.06
R_{SN} (100 μm')	0.013	0.025	0.04

substrate can be less than 0.025 ohm, which is considerably smaller than the worst possible value of 0.06 ohm and the average value of 0.05 ohm found in practice.

2.4.2.4. The P⁺ Layer (R_P)

The vertical resistance of the P⁺ layer is 0.008 ohm, assuming a surface concentration of 1.8×10^{20} cm^{-3} and a junction depth of 0.5 micron.

2.4.2.5. The Effect of the P⁺ Layer without the Metal Electrode (R_{ple})

The lateral resistance of the metal electrode that covers the P⁺ layer is negligibly small. However, the lateral resistance of the P⁺ layer not covered by the electrode should be taken into account.

The diagram expressing the effect is shown in Figure 25 and the equivalent circuit in Figure 26.

The lateral effect can be expressed by a distributed line consisting of a distributed capacitance and resistance, as shown in the figure. At frequencies sufficiently below cutoff of the distributed line, the equivalent resistance (R_{Pl}) is ⅓ of the total resistance measured from edge to edge. R_{Pl} is expressed as

$$R_{Pl} = \frac{1}{3} \cdot \rho_s \cdot \frac{a_P}{4a_{Al}} = \frac{\rho_s a_P}{12a_{Al}} \tag{2-40}$$

where ρ_s is the sheet resistivity of the P⁺ layer, a_P is the distance from the edge of the metal electrode to that of the P⁺ layer, and a_{Al} is the width of the side of the aluminum electrode. The value of a_P is assumed to be sufficiently smaller than a_{Al}.

In order to convert the effective lateral resistance (R_{Pl}) to the effective series resistance (R_{Ple}), the equivalent circuit conversion

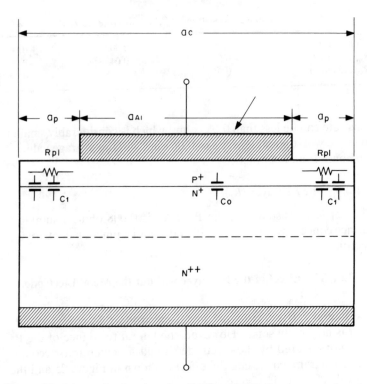

Figure 25. Schematic showing the distributed CR line representing the effect of the lateral resistance of the P^+ layer.

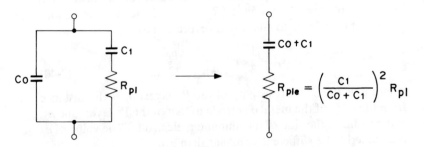

Figure 26. Equivalent circuit representing the effect of the lateral resistance of the P^+ layer.

Table 3. Equivalent lateral resistance of the P$^+$ layer

a_p (μm)	12.5	10	7.5	5
$C_1/(C_1 + C_0)$	0.234	0.190	0.144	0.097
R_{Pl} (ohms)	0.238	0.185	0.135	0.088
R_{Ple} (ohms)	0.056	0.035	0.019	0.009

shown in Figure 25 is performed. The effective series resistance (R_{Ple}) is then expressed as

$$R_{Ple} = \left(\frac{C_1}{C_0 + C_1}\right)^2 R_{Pl}$$

(2-41)

where C_0 is the capacitance under the metal electrode, and C_1 is that under the P$^+$ layer not covered with the electrode.

The effect is calculated for several cases, assuming the sheet resistivity (ρ_s) to be 40 ohms per square, as shown in Table 3. Distance a_p is 12.5 microns in the case of the 1T6 diode. If a_p is decreased to 5 microns, the effective series resistance becomes 0.009 ohm, which is much smaller than the 0.056 ohm of the 1T6 diode.

2.4.2.6. Contact Resistance between Aluminum and the P$^+$ Layer (R_{cP})

There are various reports on the specific resistance (ρ_{cAl}) of the contact between aluminum and the P-type silicon. The data reported by H. Sello is applied[14] and the following is obtained, with the assumption that the surface concentration is 1.8×10^{20} cm^{-3} and the contact area is 175 microns square.

$$\rho_{cAl} = 10^{-6} \text{ ohm-cm}$$
$$R_{cP} = 0.003 \text{ ohm}$$

2.4.2.7. Contact Resistance of the Back of the Substrate (R_{cAu})

As an estimate, the data reported by D. Shinoda on platinum silicide[15] is applied. The specific contact resistance (ρ_{cPt}) to N-type silicon with a resistivity of 0.003 ohm-cm is

$$\rho_{cPt} = 1.5 \times 10^{-5} \text{ ohm-cm}^2$$

The substrate is 400 microns square, yielding a contact resistance of gold (R_{cAu}) as

$$R_{cAu} = \frac{1.5 \times 10^{-5}}{(4 \times 10^{-2})^2} \fallingdotseq 0.009 \text{ ohm}$$

2.4.2.8. Bonding Wire (R_w)

A gold wire with a diameter of 50 microns is used for bonding the silicon chip to the outer lead. The wire has considerable resistance due to the skin effect.

The conductivity of gold is

$$\rho_{Au} = 2.3 \times 10^{-6} \text{ ohm-cm}$$

The skin depth at 470 MHz then becomes 3.5 microns. Therefore, the resistance of the gold wire with a diameter of 50 microns per unit length (1 mm) is

$$R_w = \frac{10^{-1} \times 2.3 \times 10^{-6}}{5 \times 10^{-3} \times 3.14 \times 3.5 \times 10^{-4}} = 0.042 \text{ ohm/mm}$$

The actual length is 1.25 mm, resulting in a series resistance (R_w) of

$$R_w = 0.042 \times 1.25 \fallingdotseq 0.053 \text{ ohm}$$

The series resistance was measured for experimental diodes (TX 259, lot 484-A) bonded with wires having three different diameters, that is, 25, 50, and 75 microns. Forty devices were measured and the data were averaged. The results are shown in Table 4, where a comparison is made with the calculated values.

If a wire with a diameter of 75 instead of 50 microns is used, an improvement of 0.019 ohm will be possible. The lower value for the experiment in the case of the 25-micron wire may be due to inaccuracies in the actual wire length.

2.4.2.9. The Outer Leads (R_l)

The lead material is covar, with gold plated to a thickness of 2 microns. Electrical conduction is mostly through the gold layer.

Table 4. Comparison of measured resistances with different bonding wire diameters

Wire diameter (μm)	25	50	75
Average series resistance, R_s (ohms)	0.625	0.581	0.562
Measured increase of R_s (ohms)[a]	0.039	0	−0.019
Calculated increase of R_s (ohms)[a]	0.053	0	−0.018

[a]The increase of series resistance is calculated with reference to the value of the diode bonded with a 50-micron diameter wire.

The thickness of the lead is 250 microns and the width is 0.8 mm. The parts of the leads within the plastic mold are 3.6 mm in length. The series resistance is calculated as

$$R_1 = 2.3 \times 10^{-6} \times \frac{0.36}{(0.08 + 0.025) \times 2 \times 2 \times 10^{-4}}$$
$$\cong 0.020 \text{ ohm}$$

Using silver as the plating material improves the resistance to

$$R_1 = 0.009 \text{ ohm}$$

where the skin depth at 470 MHz is calculated as 2.95 microns.

2.4.3. Summary of Contributing Factors

The factors contributing to the series resistance are summarized in Table 5, together with the means for improvement and the possible improved results.

The series resistance of the reference diode is around 0.63 to 0.75 ohm, and the improved value could be around 0.53 to 0.59 ohm, suggesting a possible improvement of 0.1 to 0.16 ohm.

The approximations may lead to some overestimations, but the obtained results exhibit general design principles for diodes with low series resistance. We can see that the nonuniformity of the epitaxial layer, both in thickness and in resistivity, is the largest contributing factor.

Table 5. Contributing factors to the series resistance and means for improvement

Factor	Value (ohms)	Means for improvement	Reduction (ohms)	Improved value (ohms)
Epitaxial layer	0.36–0.48	Improve uniformity (deviation to ½)	0.06	0.36–0.42
N⁺ layer	0.07	None	0	0.07
Substrate	0.05	Reduce thickness (100 μm), ρ (0.002 ohm-cm max.)	0.025	0.025
P⁺ layer	0.008	None	0	0.008
P⁺ layer without Al	0.056	Reduce uncovered area (12.5 to 5 μm)	0.047	0.009
Contact of P⁺ and Al	0.003	None	0	0.003
Contact of N⁺ and Au	0.009	None	0	0.009
Bonding wire	0.053	Wire with larger diameter (50 to 75 μm)	0.019	0.034
Outer leads	0.020	Au plating to Ag	0.011	0.009
Sum	0.629–0.749		0.102–0.162	0.527–0.587

2.5. Practical Considerations

2.5.1 SiH_4 Epitaxy

As indicated in the previous section, the uniformity of the epitaxial layer in terms of thickness and resistivity is very important to control the series resistance to a small value. It is important for the uniformity of the CV curve as well.

In general, the deviation of the ion-implanted profile is expected to be much smaller than the characteristics of the epitaxial layer. Therefore, controllability of epitaxial growth is of great concern in fabrication of voltage-variable capacitance diodes that require the control of the impurity profile.

During epitaxial growth, so-called autodoping is another undesirable effect. Autodoping is the substrate impurity moving into the epitaxially grown layer, either through etching of the substrate by chlorine or through thermal outdiffusion of impurities from the substrate.

As described in the previous section, the resistivity of the substrate contributes considerably to the series resistance. Substrates with a very high concentration of arsenic should be used, making the autodoping phenomenon a crucial concern. With respect to nonlinear distortion, autodoping gives the profile an opposite slope to the one derived from the nondistortion curve.

SiH_4 epitaxy does not involve chlorine and the growth temperature is around 1000°C, about 100 degrees lower than in other systems.

Autodoping has been successfully controlled, as is described in the next section. The thickness was controlled to within ± 0.1 microns within a wafer and ± 0.3 micron within a batch. The resistivity was within $\pm 7\%$ of the desired 1.0 ohm-cm.

2.5.2. Self-Aligned Structure

The diode fabricated by the basic process shown in Figure 16 has a P^+ diffused area larger than the N^+ implanted area. In other words, the edge of the P^+ layer directly faces the epitaxially grown N-type layer with much less concentration. Thus, the high breakdown voltage required is attained by the field-relieving effect of the structure.

However, the P^+ region facing the epitaxially grown N-type layer contributes very little to the capacitance change, and thus acts as a parasitic capacitance. In order to compensate for this parasitic capacitance, a larger capacitance ratio is required, resulting in a diode design yielding a higher series resistance.

To reduce the fixed capacitance of the P^+ layer to a minimum, a self-aligned structure was developed. In this structure, the implantation for the N^+ layer and the diffusion of the P^+ layer are performed through the same masking window. The structure is shown in Figure 27.

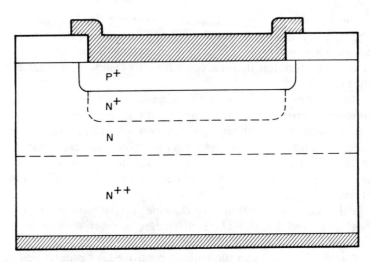

Figure 27. Cross section of a self-aligned structure.

2.5.3. Package

The requirements for the package are high reliability and low parasitic capacitance. Plastic-mold diode packages have been extensively used.[16-18]

Due to the ever increasing requirement for smaller size, packages have evolved in three generations, the newest being the smaller. They are shown in Figure 28. All three models are still in use. The leads are made of silver-plated copper.

Shipment in taped and reeled form has become increasingly common, reflecting the wide use of automatic assembly.

2.6. Diodes for VHF and UHF Tuners

2.6.1. Requirements

The general requirements for voltage-variable capacitance diodes are a high capacitance ratio and a low series resistance. However, the specifications for the diodes for VHF and UHF tuners have

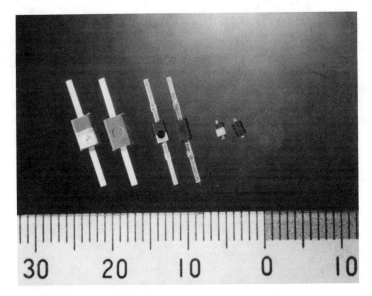

Figure 28. Packages for voltage-variable capacitance diodes.

Table 6. General requirements of voltage-variable capacitance diodes and typical values for the TX259

Characteristic	Requirement	Value (TX259)
C_2 (pF)	>13.5	15
C_{25} (pF)	<2.5	2.27
R_s (ohms at 470 MHz)	<0.6	0.52
Reverse breakdown voltage (V)	>32	35

standardized to certain values, reflecting various requirements for circuit applications and the progress of technology. The requirements are summarized in Table 6 along with the typical values obtained for a developmental device (TX259).

In this section, the fabrication process and the structure of a typical device together with the associated characteristics are described.

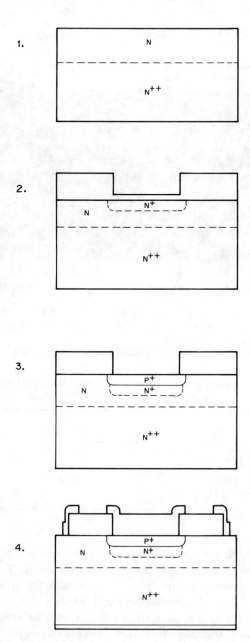

Figure 29. Fabrication process and structure of a practical diode.

2.6.2. Fabrication Process and Structure

The fabrication process and structure are shown in Figure 29. An arsenic-doped N-type substrate with a typical very low resistivity of 0.002 ohm-cm was employed.

An N-type layer of 3.0 microns with a resistivity of 1.0 ohm-cm was formed by SiH_4 epitaxy.

$^{31}P^{+2}$ ion implantation with an energy of 500 keV was performed from the direction of partial channeling through a thin SiO_2 film.

The P^+ layer was diffused to a depth of 0.5 micron.

Aluminum was evaporated for the top metal electrode, and gold was used for the back electrode.

The bonding wire was 50 mm in diameter.

2.6.3. Characteristics

The typical characteristics are summarized in Table 6. The capacitance is plotted against the reverse applied voltage in Figure 30, and the associated measure of nonlinear distortion (ξ) in Figure 31. ξ is very small, reflecting the effect of partial channeling.

The carrier-concentration profile calculated from the CV curve is shown in Figure 32. The abrupt transition of the concentration between the substrate and the epitaxial layer obtained is an advantage of SiH_4 epitaxy.

The temperature coefficient of the diode capacitance is plotted as a function of the reverse applied voltage in Figure 33.

All the characteristics should be designed and measured so that the diodes are usable within a wide ambient-temperature range. A few volts of margin are necessary for the reverse breakdown voltage at room temperature. The temperature coefficient of the capacitance at 2 volts is considerably large, mainly due to the large temperature coefficient of the diffusion potential of the junction.

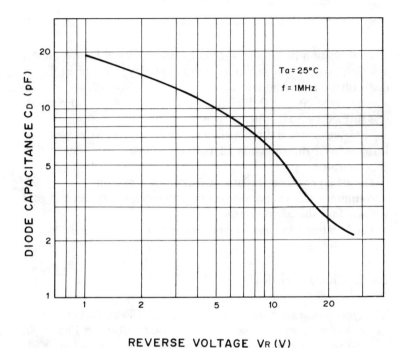

Figure 30. CV curve of the fabricated diode.

2.7. Voltage-Variable Capacitance Diodes for CATV Tuners

2.7.1. Design Considerations

CATV tuners are characterized by the large number of TV channels. In many cases, more than 100 channels must be covered. In order to cover this wide frequency range, voltage-variable capacitance diodes with large capacitance ratios are required, together with band-switch diodes that change the equivalent inductance values by acting as high-frequency electronically controllable switches. The requirement for the large capacitance change leads to a thick undepleted layer with low carrier concentration at low bias voltages,

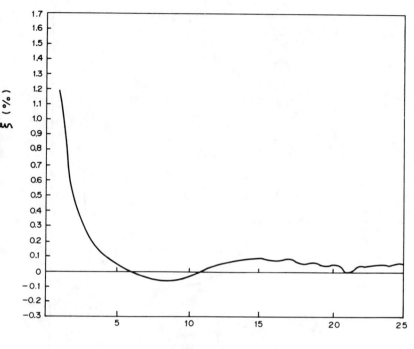

REVERSE APPLIED POTENTIAL (VOLTS)

Figure 31. Measure of nonlinear distortion calculated for the fabricated device.

thus contributing to the increase of the series resistance, which lowers the Q value of the diode. The large capacitance change also tends to increase the nonlinear distortion caused by the sharp capacitance changes. It also requires doping of impurities deep into the device.

2.7.2. Device Fabrication[19]

A device design utilizing partial channeling and a two-stage implantation scheme with different accelerating energies at each stage

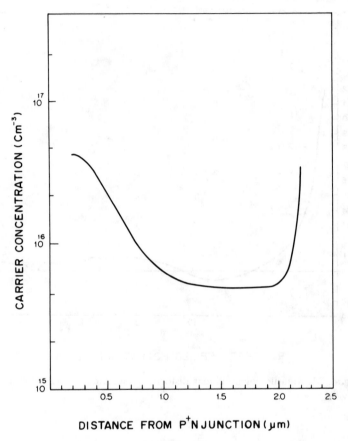

Figure 32. Carrier-concentration profile of the fabricated device.

were adopted. Partial channeling is effective in introducing impurities into deep regions, thereby decreasing nonlinear distortion.

The design target is summarized in Table 7.

The junction area of the developmental diode was 5.706×10^{-4} cm^2. The process was basically similar to that for the devices for TV tuners. The impurity profile is shown in Figure 34. The value of

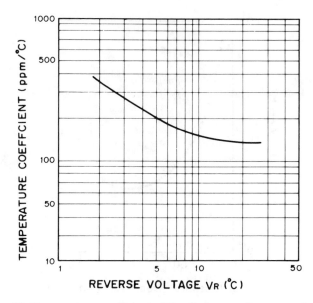

Figure 33. Temperature coefficient of the diode capacitance as a function of the reverse-bias voltage.

Table 7. The design target

C_{25}	3.2 pF (maximum)
C_2/C_{25}	10 (minimum)
r_s	0.7 ohm (maximum)

the resistance calculated from the profile is around 0.46 ohm, satisfying the design target for the resistance of the N layer.

2.7.3. Characteristics

The characteristics of the fabricated device (TX393) and those of 1T31[20] are shown in Table 8. Although the capacitance ratio of the fabricated device is larger, the series resistance is smaller in comparison with 1T31, showing the improvement in design.

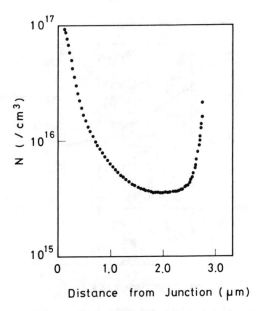

Figure 34. Impurity-concentration profile of a diode with a large capacitance ratio.

Table 8. Characteristics of the diodes

Parameter	TX393	1T31
Capacitances (pF)		
C_2	38.00	30.00
C_{25}	2.515	2.9
r_s (ohms at 470 MHz)	0.65	0.7

A typical CV curve is shown in Figure 35. The associated values of the measure of the nonlinear distortion (ξ) are shown in Figure 36. The new design shows good distortion characteristics. The values of m [$= -V \, dC/C \, dV$] are shown in Figure 37, indicating better performance of the fabricated device. The technology was basically applied to the 1T363 device (Sony), which is commercially available.[21]

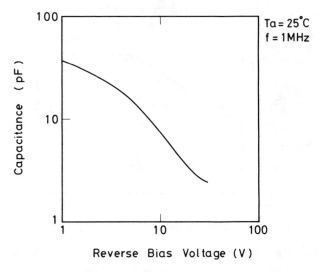

Figure 35. Capacitance of the diode as a function of the reverse-bias voltage.

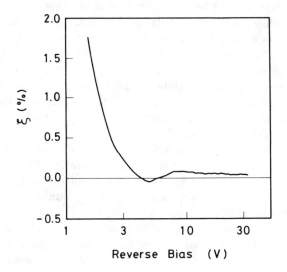

Figure 36. Measure of the nonlinear distortion ξ.

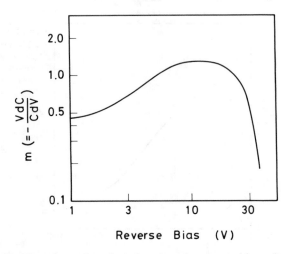

Figure 37. The values of m plotted against the reverse-bias voltage.

2.8. GaAs Voltage-Variable Capacitance Diodes

GaAs offers a much higher electron mobility than silicon. Thus, GaAs voltage-variable capacitance diodes should be superior in series resistance, and consequently in the Q value. GaAs voltage-variable capacitance diodes were extensively studied by T. Hara and his group.[9,22–24] At first, a steplike profile made by epitaxial growth was employed, as shown in Figure 6(b).[22,23] Next, multiple ion implantation of silicon was applied in order to reduce the nonlinear distortion, as indicated by its measure expressed by ξ in Equation (2-13).

The characterisics of typical diodes are exhibited in Table 9, where the characteristics of silicon diodes are listed for comparison. The diode with a steplike impurity profile has the smallest series resistance at the expense of a far larger value of ξ in comparison with other diodes.

They also reported a good correspondence of ξ and cross-modulation distortion for the two GaAs diodes shown in Figures 38 and 39.[9] The cross-modulation distortion is measured by the amplitude of the undesired signal (E_u), which causes a 1% cross

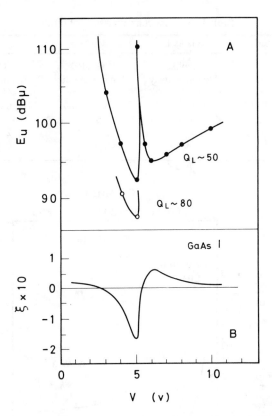

Figure 38. Cross-modulation performance of varactor-tuned circuits when double epitaxial GaAs diodes (GaAs 1) with a step profile were used. (a) Observed undesired input voltage E_u that produces 1% cross modulation as a function of dc applied bias voltage V, where the loaded Q of the circuit, Q_L, was used as a parameter. The undesired signal frequency was 12 MHz above that desired. (b) The measure of nonlinear distortion ξ as a function of applied voltage V. (© IEEE 1980, *IEEE Trans. CE*, Vol. CE-26, May 1980.)

Table 9. Typical characteristics of diodes

Parameter	GaAs 1	GaAs 3	Si
Profile	Steplike	Graded	Graded
Profile formation	Double epitaxy	Ion implantation	Ion implantation
R_s (ohms;			
470 MHz, 9 pF)	0.18	0.25 ± 0.05	0.65 ± 0.17
C_3 (pF)	12.0	11.8	11.0
C_{25} (pF)	2.0	2.0	2.0
$|\xi|$	<0.18	<0.0022	<0.002

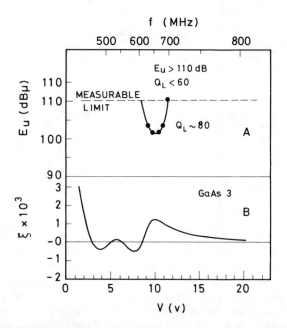

Figure 39. Cross-modulation performance of varactor-tuned circuits when low-distortion GaAs diodes (GaAs 3) fabricated by triple ion implantation were used. (a) Observed undesired input voltage E_u that produces 1% cross modulation as a function of dc applied bias voltage V, where the loaded Q of the circuit, Q_L, was used as a parameter. The undesired signal frequency was 12 MHz above that desired. (b) The measure of nonlinear distortion ξ as a function of applied voltage V. (© IEEE 1980, *IEEE Trans. CE*, Vol. CE-26, May 1980.)

Table 10. Temperature coefficients of diode capacitances at different bias voltages at 20°C and 1 MHz[22]

Applied voltage (volts)	GaAs diode (ppm/°C)	Si diode (BB105) (ppm/°C)
3	146	411
12	10	134
25	7	85

modulation to the desired signal usually set 12 MHz apart from the undesired. Therefore, the larger the undesired signal amplitude (E_u), the smaller the cross-modulation distortion.

Another advantage of GaAs diodes, the low temperature coefficients of diode capacitances in comparison with silicon diodes, was reported,[22] as shown in Table 10.

GaAs diodes are not widely used for consumer applications. The reason lies, we believe, in the immature process and material technologies in comparison with silicon. However, as the process and material technologies advances rapidly, full application of GaAs diodes will be realized.

2.9. Summary

The disadvantages of voltage-variable capacitance diodes to mechanically variable capacitors can be summarized as follows:

1. Nonlinear effects.
2. Low Q value due to the series resistance.
3. Nonuniformity due to processing-parameter deviations.

The design and the fabrication technologies for reducing these disadvantages have been investigated. They are summarized as follows:

1. A measure of nonlinear distortion effects was derived as a function of the first and second derivatives of the CV curve.
2. A CV curve that gives the zero value of the measure of nonlin-

ear distortion was then analytically derived, and the corresponding impurity profile of the hyperabrupt junction diode was also derived, providing a design guide for diodes with low-distortion effects.

3. As a means for implanting impurities deep into silicon, which is required for low distortion, $^{31}P^{+2}$ implantation from the direction of partial channeling was investigated. It was found that implantation through a thin SiO_2 film is effective in achieving uniformity of profile.

4. Various factors contributing to the series resistance were analyzed in order to establish a design principle for diodes with low series resistance. It was shown that the nonuniformity of the epitaxial layer is a very large contributing factor.

5. It was shown that SiH_4 epitaxy is an effective technique for realizing a uniform profile without autodoping effects.

6. Actual sample devices for the VHF and UHF tuners were fabricated and indicated a very low series resistance of 0.56 ohm at 470 MHz, with associated capacitances of 15 and 2.27 pF at biases of 2 and 25 volts.

7. Diodes for CATV tuners, which require much larger capacitance ratios, were designed and fabricated. The series resistance was 0.65 ohm at 470 MHz with associated capacitances of 38 and 2.515 pF at 2 and 25 volts, respectively. The measure of the nonlinear distortion was low throughout the operation range.

8. GaAs voltage-variable capacitance diodes indicate very low series resistances as well as low temperature coefficients of the capacitances.

References

1. M. H. Norwood and E. Shatz. Voltage Variable Capacitor Tuning: A Review. *Proc. IEEE* **56,** 5(1968): 788–798.
2. Y. Ninomiya. Crossmodulations in Variable-Capacitance Diode Tuners, *NHK Tech. J.* **24,** 1(1972): 44–58.
3. S. Watanabe. Japanese patent, filed, No. S55-16461.
4. S. Hilliker. Avoiding Pitfalls in Varactor Circuit Design. *IEEE Trans. Consumer Electron.* (August 1976): 195–202.

5. S. Watanabe. Nonlinear Distortion of Voltage Variable Capacitance Diodes. *IECE Technical Report*, No. SSD76-36, pp. 1–9, 1976.
6. S. Watanabe, H. Yamoto, T. Aoki, H. Kubota, and Y. Hayashi. Low Voltage Variable Capacitor Made by Ion Implantation and Silane Epitaxy. *Denshi Tokyo* **15**, (1976): 53–55.
7. R. G. Meyer and M. L. Stephens. Distortion in Variable-Capacitance Diodes. *IEEE J. Solid-State Circ.* **SC-10**, 1(1975): 47–54.
8. J. H. Mulligan, Jr., and C. A. Paludi, Jr. Varactor Tuning as a Source of Intermodulation in RF Amplifiers. *IEEE Trans. Electromag. Compat.* **EMC-25**, 4(1983): 412–421.
9. T. Hara, I. Niikura, and N. Toyoda. TV Tuners Using Low Loss and Low Distortion Varactor Diodes. *IEEE Trans. Consumer Electron.* **CE-26** (1980): 172–179.
10. R. A. Moline and G. F. Foxhall. Ion-Implanted Hyperabrupt Junction Voltage-Variable Capacitors. *IEEE Trans. Electron Dev.* **ED-19**, 2(1972): 267–273.
11. А. М. МАЛЛЯР, Радиотехника, т. **22**, 2(1967): 16–23.
12. H. Kubota, Y. Hayashi, N. Okazaki, and S. Watanabe. A P^{+2} Implanted Hyperabrupt Junction Voltage Variable Capacitor, in *Extended Abstracts of the 9th Symposium on Semiconductors and Integrated Circuits Technology*, The Electrochemical Society of Japan, pp. 50–55, 1976.
13. V. G. K. Reddi and J. D. Sansbury. "Channeling and Dechanneling of Ion-Implanted Phosphorous in Silicon." *J. Appl. Phys.* **44**, 7(1973): 2951–2963.
14. H. Sello. Ohmic Contacts and Integrated Circuits, in B. Schwartz, ed., *Ohmic Contacts to Semiconductors,* pp. 31–47, The Electrochemical Society (1969).
15. D. Shinoda. Ohmic Contacts to Silicon Using Evaporated Metal Silicides, in B. Schwartz, ed., *Ohmic Contacts to Semiconductors,* pp. 200–213, The Electrochemical Society (1969).
16. Sony Discrete Device Data for 1T25.
17. Sony Discrete Device Data for 1T32.
18. Sony Discrete Device Data for 1T362.
19. H. Sone. Private communication.
20. Sony Discrete Device Data for 1T31.
21. Sony Discrete Device Data for 1T363.
22. T. Hara, I. Niikura, T. Hozuki, N. Toyoda, and M. Mihara. GaAs Varactor Tuners for UHF TV. *IEEE Trans. Consumer Electron.* **CE-23**, 4(1977): 433–439.
23. T. Hara. I. Niikura, N. Toyoda, and M. Mihara. High-Q GaAs

Varactor Diodes. *IEEE Trans. Electron Dev.* **ED-25,** 5(1978): 501–506.

24. T. Hara, N. Toyoda, and I. Niikura. Resistance of GaAs Varactor Diodes for TV Tuners. *IEEE Trans. Consumer Electron.* **CE-26** (1980): 729–736.

CHAPTER 3

Band-Switch Diodes

Abstract

Design and the fabrication process for band-switch diodes with high Q values are described.

3.1. Introduction

3.1.1. Requirements of Band-Switch Diodes

The frequency coverage of the voltage-variable capacitance diode is limited by circuit matching and the series resistance associated with the large capacitance ratio.

However, the VHF television band often consists of a separate lower band and a higher band. Also, due to the wide acceptance of CATV systems, especially in the United States, much wider frequency coverage was required for the tuner. Tuners for CATV reception today have more than 100 channels to cover.

For resonant circuits that cover a wide frequency range, the inductances are switched by a band-switch diode, as shown in Figure 40. When the band-switch diode is in forward bias, inductance L_2 is short-circuited and the inductance is effectively L_1. When it is in reverse bias, the resonant circuit has an inductance that is the sum of L_1 and L_2.

The resistance of the diode in forward bias is called the on-resistance and behaves as a parasitic resistance that lowers the Q value of the circuit, hence degrading the selectivity and the insertion loss. The capacitance in reverse bias acts as a parasitic capacitance and degrades inductance L_2.

Therefore, the product of the on-resistance and the capacitance in reverse bias is regarded as the figure of merit of the diode. The value of the capacitance is usually set to be a little less than 1 pF

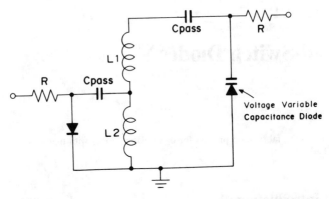

Figure 40. Resonant circuit containing a band-switch diode. When the diode is reverse biased, the inductance is the sum of L_1 and L_2. When it is forward biased, L_2 is effectively shunted, and the inductance is L_1.

considering the requirement of the circuit; the on-resistance then becomes the parameter to evaluate the performance of the diode. The on-resistance is usually measured and specified at 100 MHz.

3.1.2. Basic Structure of Band-Switch Diodes

The basic structure of the band-switch diode is shown in Figure 41. The package is plastic mold, usually the same as that for the voltage-variable capacitance diode.

The diode requires an intrinsic layer, which is usually realized by an N-type layer with a very low impurity concentration. This layer becomes completely depleted for reverse bias and acts as a capacitor. For forward bias, the carriers are stored in this region and the resistance becomes very low, making the diode act as a turned-on switch.

The intrinsic layer is formed by epitaxial growth. The impurity concentration of the epitaxial layer should be as low as possible in order to make the diode act as an ideal capacitance for reverse bias. An abrupt transition from the substrate with a very high impurity concentration to the epitaxially grown layer is also essential.

The lifetime of the carrier is another important parameter. The

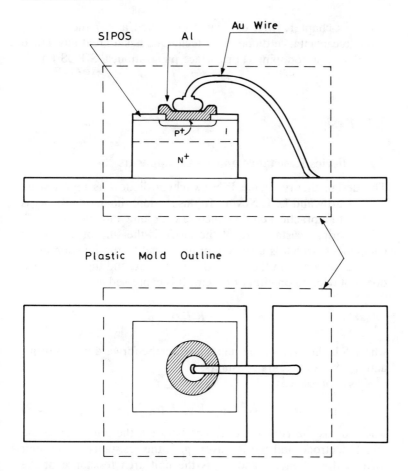

Figure 41. The basic structure of the band-switch diode. It is basically a PIN structure.

lifetime is influenced by the fabrication process as well as by the device structure. Therefore, it is practical to develop a process yielding a diode with a lifetime as long as possible.

The surface of the intrinsic layer is subject to inversion because the impurity concentration is very low. Reliability thus becomes another important issue.

In this chapter, SiH_4 epitaxy is employed as for the voltage-variable capacitance diode. This technique has a great advantage in reducing autodoping. For surface passivation, a SIPOS film is used.[1,2]

3.2. Design

3.2.1. Series Resistance and Its Components

The design theory of the PIN switching diode was reported by R. C. Curby and L. J. Nevin.[3] In this section, the design theory is applied to a practical diode structure with some modifications.

The series resistance R_s of the diode is the sum of the ohmic component, which is inversely proportional to the junction area, one that depends on the forward-bias current, and the rest, which does not depend on either of them. It is expressed as

$$R_s = \frac{K_1}{S} + K_2 I_F^{-2/3} + K_3 \tag{3-1}$$

where S is the area of the junction, I_F is the forward-bias current, and K_1, K_2, and K_3 are constants.

K_1 is expressed as

$$K_1 = \rho_{cP} + \rho_P + \rho_{N1} \tag{3-2}$$

where ρ_{cP} is the contact resistance between the unit area of the metal electrode and the P^+ layer, ρ_P is the resistance of the unit area of the P^+ layer, and ρ_{N1} is the unit area resistance of the portion of the substrate that can be regarded as inversely proportional to the junction area.

The resistance R_I, which depends on the current, is expressed as

$$R_I = \frac{w^2}{2\bar{\mu} I_F \tau_{eff}} \tag{3-3}$$

where w is the thickness of the intrinsic layer, $\bar{\mu}$ is the average mobility of holes and electrons, and τ_{eff} is the effective carrier lifetime.

According to R. C. Curby and L. J. Nevin, τ_{eff} has a current dependence as follows:

$$\tau_{\text{eff}} = \tau_0 \frac{w}{w_0} \left(\frac{J_0}{J}\right)^{\frac{1}{3}}$$

$$(3\text{-}4)$$

where τ_0 is the reference carrier lifetime at a current density of J_0 with an intrinsic layer thickness of w_0.

The dependence was confirmed in the structure shown in Figure 40 as is indicated in Figure 42, where t_{rr} of two diodes are plotted against the forward current.

If the intrinsic layer is completely depleted for reverse bias, capacitance (C_j) is

$$C_j = \frac{\varepsilon S}{w}$$

$$(3\text{-}5)$$

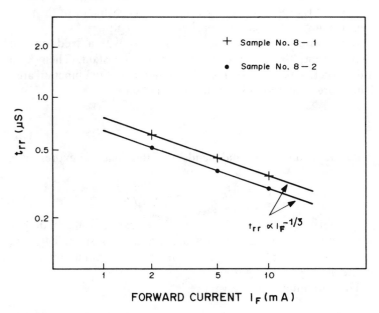

Figure 42. t_{rr} is plotted as a function of forward current I_F, showing that t_{rr} is proportional to $I_F^{-\frac{1}{3}}$.

Then K_2 is expressed as

$$K_2 = \frac{\varepsilon w_0 S^{2/3}}{2\bar{\mu} C_j \tau_0 J_0^{1/3}} \tag{3-6}$$

As can be seen from Figure 40, there are a number of contributing factors that do not depend on either the area or the current. Assuming a constant die size, the resistance of the bonding wire (R_w) and the outer leads (R_1), the contact resistance of the back of the substrate (R_{cN}), and the resistance of the substrate (R_{N2}) can be regarded as constant. Then K_3 is

$$K_3 = R_w + R_1 + R_{cN} + R_{N2} \tag{3-7}$$

3.2.2. Optimization of the Structure[3]

The optimum capacitance can be regarded as fixed because the value is determined from circuit considerations. Usually, the desirable value is around 0.8 pF, including the capacitance of the package, leading to a junction capacitance of 0.6 pF.

If the series resistance is to be considered with a fixed forward-bias current, the current also becomes a constant. Then K_2 in Equation (3-6) can be expressed as a function of the junction area. Therefore, Equation (3-1) is transformed into

$$R_s = \frac{K_1}{S} + K_2' S^{2/3} + K_3 \tag{3-8}$$

where K_2' is a constant independent of the area and expressed as

$$K_2' = K_2 I_F^{-2/3} S^{-2/3} = \frac{\varepsilon w_0}{2\bar{\mu} C_j \tau_0 J_0^{1/3}} I_F^{-2/3} \tag{3-9}$$

The series resistance in Equation (3-8) can then be minimized with reference to the junction area, and, correspondingly, the thickness of the intrinsic layer and hence the optimum dimension of the diode can be determined.

The minimum series resistance R_{sm} is

$$R_{sm} = \left(\tfrac{2}{3} K_1^{2/3} K_2'\right)^{3/5} + \left(\tfrac{3}{2} K_1 K_2'^{3/2}\right)^{2/5} + K_3 \tag{3-10}$$

when the junction area S_m is expressed as

$$S_m = \left(\frac{3K_1}{2K_2'}\right)^{0.6}$$

(3-11)

The corresponding thickness of the intrinsic layer is

$$w_m = \frac{\varepsilon S_m}{C_j} = \frac{\varepsilon}{C_j}\left(\frac{3K_1}{2K_2'}\right)^{0.6}$$

(3-12)

In practical applications, the current is a problem, since the lower the current, the better. For the optimization of the device-fabrication process, lifetime τ_0 is the parameter of most interest. The minimum series resistance R_{sm} in Equation (3-10) is expressed as a function of I_F and τ_0:

$$R_{sm} = AC_j^{-3/5}I_F^{-2/5}\tau_0^{-3/5} + K_3$$

(3-13)

where A is a constant and expressed as

$$A = [(2/3)^{3/5} + (3/2)^{2/5}]K_1^{2/5}(\varepsilon W_0/2\mu)^{3/5}J_0^{-1/5}$$

(3-14)

As can be seen in Equation (3-13), τ_0 is the critical parameter for a fixed current. The series resistance decreases if the lifetime is increased. The improvement of the lifetime corresponds to a decrease of K_2 in Equation (3-1) and to an increase of optimum junction area S_m in Equation (3-11), thus resulting in a thicker intrinsic layer.

3.2.3. Approximation of Series Resistance (K_3)

3.2.3.1. The Substrate

The contribution from the low-resistivity substrate depends both on the junction area and the structural dimension. In order to find the optimum design, it is convenient to separate the two factors, as shown in Figure 43.

The substrate is divided into two plates, one with the area of the junction and the other with the area of the substrate. The series resistance is assumed to be the sum of the resistances of the two plates:

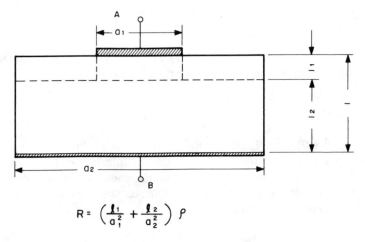

$$R = \left(\frac{l_1}{a_1^2} + \frac{l_2}{a_2^2} \right) \rho$$

Figure 43. Model for calculating the approximate resistance of the substrate. The substrate is divided into two plates with thicknesses of l_1 and l_2 proportional to a_1 and a_2, respectively, the side lengths of the top and bottom metal electrodes. The resistivity of the substrate is designated as ρ.

$$\begin{aligned} R &= \rho \frac{l_1}{a_1^2} + \rho \frac{l_2}{a_2^2} \\ &= R_{N1} + R_{N2} \end{aligned} \tag{3-15}$$

where ρ is the resistivity of the substrate, a_1 and a_2 are the side lengths of the junction and the substrate, respectively, l_1 and l_2 are the thicknesses of the two plates, and R_{N1} and R_{N2} are the resistances dependent and independent, respectively, of the junction area.

In the estimate used in this section, l_1 and l_2 are assumed to be proportional to the side lengths of the junction and the substrate, respectively. In the calculation of the optimum junction area, a_1 is assumed to be variable and l_1 to be constant.

In the case where the substrate is 400 microns square, the junction is 100 microns square, and the thickness of the substrate is 150 microns, l_1 and l_2 are 30 and 120 microns, respectively. When the resistivity of the substrate (ρ) is 0.002 ohm-cm, then R_{N1} and R_{N2} become

$$R_{N1} = 0.075 \text{ ohm}$$
$$R_{N2} = 0.019 \text{ ohm}$$

Although it may be theoretically slightly inaccurate, the approximation will not cause a substantial error on the condition that the variation of the junction area is sufficiently small. Therefore, the approximation may be justified for the optimum design for the junction area and, consequently, for the thickness of the intrinsic layer.

3.2.3.2. Other Factors

The contributing factors to series resistance K_3 that are independent of the junction area or the current have to be calculated in the way as was the voltage-variable capacitance diode in the previous chapter.

Now the resistance of the gold bonding wire at 100 MHz (R_w) was calculated as shown in Table 11. The length of the wire was assumed to be 1.25 mm. A wire with a diameter of 50 microns contributes 0.012 ohm, which is 0.012 ohm less than that of a wire with a diameter of 25 microns.

As for the contact resistance at the back of the substrate (R_{cN}), the data for platinum silicide is employed instead of the contact resistance of gold and N-type silicon with a resistivity of 0.002 ohm-cm,[4] yielding a characteristic contact resistance of 1×10^{-5} ohm-cm^2. Because the substrate is 400 microns square, the following contact resistance is obtained:

$$R_{cN} = 0.006 \text{ ohm}$$

The leads are made of silver-plated copper as thick as 5 microns. The resistance of the leads within the plastic-mold outline (R_1) is then calculated as

$$R_1 = 0.004 \text{ ohm}$$

Table 11. Resistance of gold bonding wires

Diameter (μm)	25	50	75
R_w (ohms at 100 MHz)	0.024	0.012	0.008

In the case of a bonding wire 25 microns in diameter, K_3 is estimated to be

$$K_3 = R_m + R_1 + R_{cN} = 0.034 \text{ ohm}$$

3.3. Fabrication Process

3.3.1. Requirements

The requirements for the fabrication process are as follows:

1. An epitaxial layer with a low impurity concentration and an abrupt interface to the substrate.
2. Surface passivation for a low-concentration N^- layer.
3. Processes that result in a long carrier lifetime.

3.3.2. SiH₄ Epitaxy

SiH_4 epitaxy has distinct advantages over other systems in obtaining an abrupt interface to the substrate with a high concentration of arsenic because the growth temperature is about 100 degrees lower and the source materials do not contain chlorine.

Because the source materials are given in a gas, the system has an advantage in improving the purity, contributing to the high resistivity of the epitaxial layer.

The carrier-concentration profile obtained by the SiH_4 epitaxy is shown in Figure 44, where the low concentration and the abrupt interface can be observed.

3.3.3. SIPOS Film Passivation

SIPOS (Semi-Insulating POly-Silicon) is a polysilicon doped with oxygen and is known as a good surface-passivation material.[1,2] The passivation capability of SIPOS is attributed to the high concentration of traps that decreases the electric field created by the charges accumulated on the surface of the film.

The band-switch diode is continuously subjected to a reverse-bias condition, as is also the case for the voltage-variable capacitance

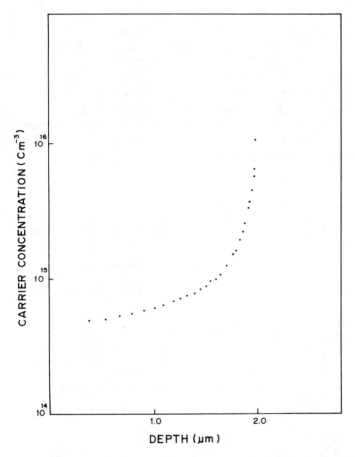

Figure 44. Carrier-concentration profile of the N^- layer grown by SiH_4 epitaxy.

diode. The interface between the surface of the passivation film and the plastic-mold resin becomes filled with negative charges. Then inversion occurs at the surface of the N^- layer, giving rise to a reverse leakage current.

The disadvantage of SIPOS film is that it has a larger reverse current than the SiO_2 passivation film. However, the requirement

for a low reverse current is not so severe for the band-switch diode
as it is for the voltage-variable capacitance diodes. The potential
drop through the high biasing resistor due to the leakage current
directly influences the resonant frequency.

In Figure 45, the reverse leakage currents are compared for the
cases of the SIPOS film and the double layers of SiO_2 and Si_3N_4.
Although the latter exhibits values below 1 nA, the former indi-
cates values below 100 nA, corresponding to a potential difference
of 1 mV across a resistor of 100 kilohms. This value can be regarded as negligible for the purpose of switching.

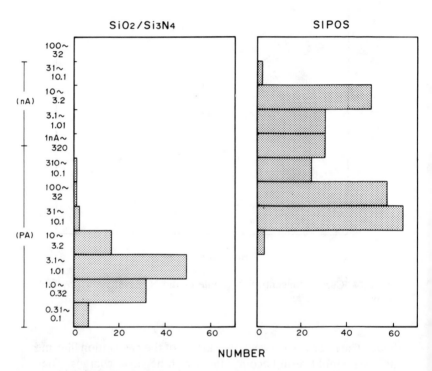

Figure 45. Distribution of the reverse leakage current in diodes passivated
with the double layer of SiO_2 and Si_3N_4 and SIPOS film. The reverse bias
applied was 28 volts.

3.3.4. Improvement of Carrier Lifetime

3.3.4.1. Carrier Lifetime and t_{rr}

Carrier lifetime is often estimated by measuring the value of t_{rr}. The value of t_{rr} is the fall time of the transient current after the forward-biased diode is switched off and short-circuited. The measuring points are 90% and 10% of the transient value.

The value of t_{rr} and the lifetime τ in the previous section are related as follows, and, therefore, t_{rr} is the direct measure of the carrier lifetime.

$$t_{rr} \doteqdot 2.2\tau \qquad (3\text{-}16)$$

3.3.4.2. Carrier Lifetime at Each Process Step

The carrier lifetime measured at each process step is plotted in Figure 46. It can be seen that the damage caused by the charged particles during plasma etching and evaporation by electron-beam heating degrades the lifetime. The damage can be annealed out by an appropriate thermal process.

3.3.4.3. Lifetime and Passivation

The recovery of t_{rr} in the intermediate process is compared for the cases of the passivation films of SIPOS and the double-layer structure of SiO_2 and Si_3N_4 in Figure 47. Note that the device with the SIPOS film exhibits a better recovery and a longer lifetime. The difference may be due to the dense structure of Si_3N_4 film, which does not allow hydrogen to reach the surface of the silicon, where damage exists to be annealed.

3.3.4.4. Lifetime and the Junction Depth of the P⁺ Region

In Figure 48, the improvement of t_{rr} with an increase in the junction depth is plotted. The reason for the increase of t_{rr} with the junction depth may be attributed to the shallow depth of the damage caused by charged particles.

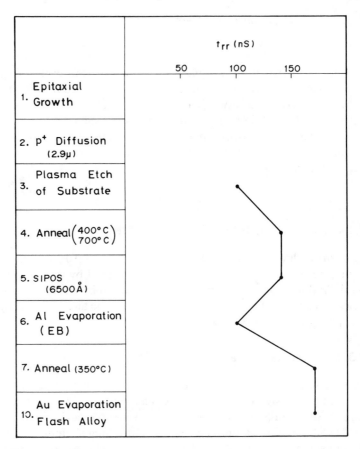

Figure 46. Fabrication process steps and t_{rr} measured after each step.

3.4. Devices and Characteristics

3.4.1. Structure

Three experimental models were fabricated and compared, namely, TX222, TX265-1, and TX265-2. The fabrication process was basically the same as that for the voltage-variable capacitance diode,

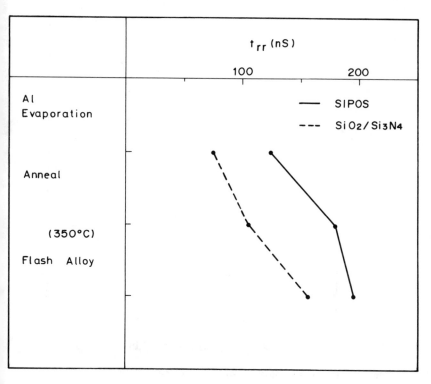

Figure 47. Comparison of the recovery of t_{rr} in the anneal and alloy processes compared between diodes passivated with a double layer of Si_3N_4 and SIPOS film.

except that there was no ion implantation. This N-type substrate with a resistivity as low as 0.002 ohm-cm was employed.

The area of the junction is smaller for TX222, and hence the smaller thickness of the intrinsic layer. The junction area of model TX222 is 10,000 square microns with a square shape. The junction area of both TX265-1 and TX265-2 is 20,000 square microns, having a round shape of 160 microns in diameter.

The depth of the junction is 0.7 micron for TX222 and 2.9 microns each for TX265-1 and TX265-2. The difference between

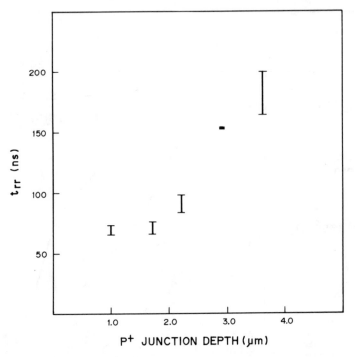

Figure 48. Improvement of t_{rr} with an increase of the P+ junction depth.

TX265-1 and TX265-2 is the impurity concentration of the P+ layer, the former being 80 ohms per square and the latter 16.3.

The package is the same as for the voltage-variable capacitance diode. The bonding wire is 25 microns in diameter for TX222 and 50 microns for both TX265-1 and TX265-2.

The structural parameters are summarized in Table 12 together with the electrical characterisics.

3.4.2. Characteristics

3.4.2.1. The Measuring Method

The series resistance (R_s) at a frequency of 100 MHz was measured in the form of admittances by a high-frequency admittance meter,

Table 12. Structural parameters and characteristics

Parameter	TX222	TX265-1	TX265-2
Junction area (10^{-4} cm^2)	1.0	2.0	2.0
Thickness of intrinsic layer (μm)	1.89	3.54	3.54
Sheet resistance of P$^+$ layer (ohm/$\square$)	40	16.3	80
Junction depth (μm)	0.7	2.9	2.9
Diameter of bonding wire (μm)	25	50	50
C_1/C_{10} (at 1 MHz)[a]	1.2	1.3	1.3
R_s (ohms) ($I_F = 2$ mA, 100 MHz)	0.66	0.46	0.60
R_s (ohms) ($I_F = 10$ mA, 100 MHz)	0.41	0.22	0.30
K_1 ($\times 10^{-5}$ ohm-cm^2)	1.97	1.97	1.97
K_2'	74.3	32.1	46.8
K_3 (ohms)	0.053	0.041	0.041
t_{rr} (ns)	$\approx$60	190–200	130–150
τ_0 (ns)	22.3	113.9	78.3

[a]C_1 and C_{10} designate capacitances for reverse-bias voltages of 1 and 10 volts, respectively.

HP Model 250B. The measured admittance was then transformed into the impedance, thus yielding the series resistance. The capacitance for the reverse bias was measured at 1 MHz by a bridge circuit, Boonton Electronics Model 75D.

3.4.2.2. Diodes with Small Junction Areas

The series resistance (R_s) of the TX222 is plotted as a function of forward-bias current (I_F) in Figure 49. From Equation (3-9), the term dependent on I_F is proportional to $I_F^{2/3}$. In Figure 49, a constant amount (0.25 ohm) was subtracted from the measured values, yielding a plot proportional to $I_F^{2/3}$. Therefore, the constant value of 0.25 ohm corresponds to the terms not dependent on I_F in Equation (3-1).

The capacitance for reverse bias is plotted in Figure 50. The curve is extremely flat, reflecting the low impurity concentration and the abrupt interface of the epitaxial layer.

The thickness of the intrinsic layer (w_0) is then calculated from the capacitance to be 1.89 microns.

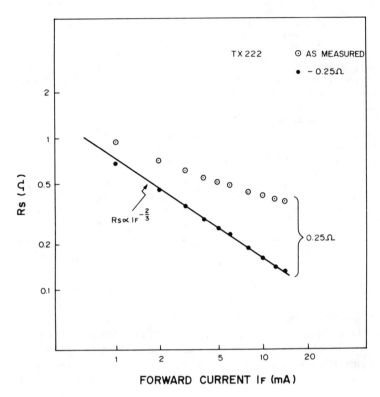

Figure 49. Series resistance of diode TX222 as a function of forward current I_F. The resistance values, after subtracting a constant value of 0.25 ohm, show a dependence of $I_F^{-2/3}$, theoretically predicted by Equation (3-9).

The term dependent on I_F ($K_2 I_F^{-2/3}$) expressed in Equation (3-8) is calculated from the figure to be 0.16 ohm at a current of 10 mA. Because the junction area is 10^{-4} cm^2, the corresponding reference current density (J_0) is calculated to be 10^2 A/cm^2. If the averaged mobility is assumed to be 500 cm^2/V-s, the lifetime (τ_0) is calculated to be 22.3 ns.

The measured t_{rr} was 60 ns, and shows reasonable agreement with the calculations according to Equation (3-15).

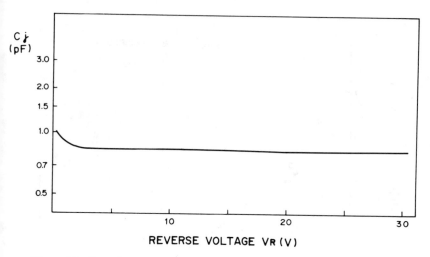

Figure 50. Capacitance of diode TX222 as a function of applied reverse voltage.

3.4.2.3. Other Parameters

The resistance component independent of current is 0.25 ohm. By subtracting K_3 ($= R_1 + R_w + R_{cN} + R_{N2}$), which was calculated in the previous section, the term inversely proportional to the area is 0.197 ohm. K_1 and K'_2 then are

$$K_1 = (0.25 - 0.053)S = 1.97 \times 10^{-5}$$
$$K'_2 = 0.16 \times S^{-\frac{2}{3}} \quad\quad = 74.3$$

3.4.2.4. Diodes with Large Junction Areas

The series resistance against the forward-bias current was plotted for model TX265-1 in Figure 51 and TX265-2 in Figure 52.

The value of 0.22 ohm at a current of 10 mA is comparable to the lowest ever reported. The values reported by R. C. Curby and L. J. Nevin were around 0.15 to 0.25 ohm with a maximum junction capacitance of 0.8 pF.

In Figure 51, a constant resistance of 0.11 ohm was subtracted

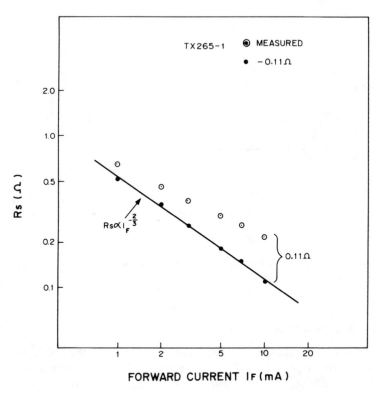

Figure 51. Series resistance of diode TX265-1 as a function of forward current I_F. The resistance values, after subtracting a constant value of 0.11 ohm, show a dependence of $I_F^{-2/3}$, theoretically predicted by Equation (3-9).

from the measured values so that the rest might exhibit proportionality to $I_F^{-2/3}$. In Figure 52, the amount to be subtracted was 0.14 ohm.

The difference in the two models may be due to the additional resistance of the P⁺ layer of the latter, which has less impurity concentration.

The bonding wire with a diameter of 50 microns resulted in a smaller value of K_3, 0.041 ohms. Parameters K_1 and K_2' are then calculated as shown in Table 12. The calculated lifetime was compared with the measured t_{rr} and the agreement was reasonable for

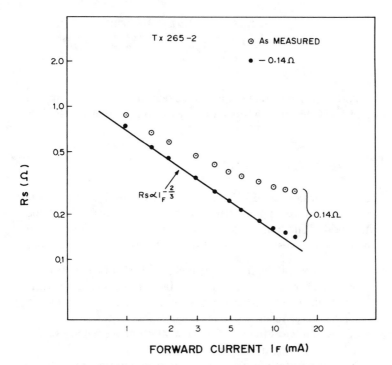

Figure 52. Series resistance of diode TX265-2 as a function of forward current I_F. The resistance values, after subtracting a constant value of 0.14 ohm, show a dependence of $I_F^{-2/3}$, theoretically predicted by Equation (3-9).

both devices. The values of the lifetime of both devices are much larger than the one with the smaller junction area, reflecting the deep junction depth.

The flatness of capacitance against the reverse bias is slightly worse than that of the TX222, probably due to the graded impurity profile as a result of the longer diffusion time, but it is still very good.

3.4.2.5. The Optimum Structure

The optimum structure was calculated from the values obtained for device TX265-2 by Equations (3-11) and (3-12), yielding dimensions close to those of TX265-1 and TX265-2.

$$S_m = (3K_1/2K_2') \fallingdotseq 1.91 \times 10^{-4} \text{ cm}^2$$
$$w_m \fallingdotseq 3.4 \text{ } \mu\text{m}$$

3.4.2.6. Reliability

The diodes were tested under a reverse bias of 28 volts at a temperature of 100°C for 1000 hours and they exhibited excellent stability.

3.5. Summary

The design principles of PIN diodes reported by R. C. Curby and L. J. Nevin were applied with some modification to the band-switch diode. The fabrication technology for diodes with high performance was investigated, resulting in the following conclusions.

1. An N-type epitaxially grown layer with a very low impurity concentration and an abrupt interface to the substrate with a very high arsenic concentration was obtained by employing SiH_4 epitaxy, resulting in an ideal capacitor for reverse bias.
2. SIPOS film was effective for passivation of the silicon surface with a very low carrier concentration.
3. The carrier lifetime was investigated in conjunction with the process parameters and it was found that a deep junction depth and appropriate annealing are effective in improving the lifetime.
4. A high-frequency series resistance of 0.22 ohm was obtained at a bias of 10 mA at a frequency of 100 MHz. The value is comparable to the best data ($R_s = 0.15–0.25$ ohm) reported by R. C. Curby and L. J. Nevin.

References

1. T. Matsushita, T. Aoki, T. Ohtsu, H. Yamamoto, H. Hayashi, M. Okayama, and Y. Kawana. Highly Reliable High Voltage Transistors by Use of the SIPOS Process. *IEEE Trans. Electron Dev.* **ED-23**, 8(1976): 826.
2. H. Hayashi, T. Mamine, and T. Matsushita. A High Power Gate-

Controlled Switch (GCS) Using New Lifetime Control Method. *IEEE Trans. Electron Dev.* **ED-28,** 3(1981): 246.

3. R. C. Curby and L. J. Nevin. "Low Resistance, Low Bias Current PIN Diodes. *Digest of Papers of the 1976 International Solid-State Circuit Conference.*

4. D. Shinoda. Ohmic Contacts to Silicon Using Evaporated Metal Silicides, in B. Schwartz, ed., *Ohmic Contacts to Semiconductors,* pp. 200–213, The Electrochemical Society (1969).

CHAPTER 4.

Dual-Gate FETs

Abstract

The analysis, design, and fabrication process of dual-gate MOSFETs and GaAs MESFETs with low distortion in the gain-control mode are described.

4.1. Introduction

4.1.1. Applications of Dual-Gate FETs

In comparison with the bipolar transistor, the FET has basically a square-law transfer characteristic and exhibits smaller third-order distortions such as cross modulation and intermodulation.

Dual-gate FETs have been widely used for the front-end amplifier of tuners because of their small feedback capacitance and their superior characteristics of gain control.

The dual-gate FET is a cascode connection of two FETs, namely, the first being of a grounded source structure and the second the grounded gate. The feedback capacitance is reduced by the voltage gain of the second FET.

In the gain-control mode of operation, the input and output impedances of the device exhibit reasonably small changes. The cross-modulation distortion stays comparatively small over the entire gain-control range.

For these reasons, dual-gate FETs are extensively used for the first stage of tuners, as well as for mixers in the VHF and UHF ranges.

4.1.2. Dual-Gate MOSFETs

Dual-gate MOSFETs were first researched in 1965 to 1970, most extensively by RCA.[1-5] In Japan, the first application of the device to VHF tuners for consumer television sets was realized by Sony in 1969 using the model 3SK37, which contained a separate chip of back-to-back diodes for the protection of the insulated gates from electric surges.

As the UHF band was opened for consumer applications, the dual-gate MOSFET for UHF tuners was also commercialized by the author's group as 3SK46 and 3SK48 in 1972 and 1973, respectively. They contained on-chip protection diodes.[6–8]

A device applicable to both VHF and UHF with grooved gates was introduced by the group including the author in 1978.[9] The structure of the self-aligned polysilicon or molybdenum gate, developed for LSIs, was employed later by various groups[10–13] and the development of this structure is still actively pursued.

4.1.3. Dual-Gate MESFETs

In the UHF band, the dual-gate MOSFET exhibits a much larger noise figure (usually about 3 dB) than in the VHF band due to the limitation determined by the relatively low electron mobility in silicon. On the other hand, the received signal amplitude is expected to be smaller as the propagation of UHF signals is strongly affected by objects in the direct line of sight.

In order to improve the noise figure of UHF tuners, dual-gate GaAs MESFETs were researched.[14–22] Although they had superior noise figures, the cross-modulation characteristics were inferior to the MOSFET. Therefore, the improvement of cross modulation has been a major issue for dual-gate GaAs MESFETs.

4.1.4. Electronic Tuners and Low-Distortion Dual-Gate FETs

In electronic tuners, the degradation of the insertion loss and the cutoff of interfering signals due to the relatively low Q value of the tuning diodes are inevitable. In order to compensate for these disadvantages, not only a low noise figure but also improved cross-modulation characteristics are required of dual-gate FETs.

However, analytical studies on third-order distortion in the gain-reduction operation mode of dual-gate FETs have not been reported so far.

In this chapter, an analysis is performed based on a simple model of dual-gate FETs, in order to draw design principles for low-distortion devices.

A MOSFET structure with grooved gates and deep ion implantation below the second gate is investigated as a possible design to meet the requirements.

The design principle is applied to the design of a dual-gate GaAs MESFET, specifically in the lengths of both gates, yielding data confirming the analysis.

4.2. Theory of Distortion in the Gain-Control Mode

4.2.1. Operation Characteristics of Dual-Gate MOSFETs

In this section, the static transfer characteristics of a dual-gate MOSFET is analyzed based on a simple model.[23] The analysis is basically applicable to dual-gate GaAs MESFETs. The cross section of a dual-gate MOSFET is shown in Figure 53 with biasing circuitry and active and parasitic capacitances.

The characteristic of the first FET in the triode region is as follows (subscript 1 designates the parameters associated with the first FET):

$$I_{D1} = \frac{\varepsilon_1 \mu_1 W_1}{2 L_1 t_{ox1}} [(V_{G1S} - V_{P1})^2 - (V_{G1S} - V_{P1} - V_M)^2]$$

$$= \frac{\beta_1}{2} [V_1^2 - (V_1 - V_M)^2] \tag{4-1}$$

$$\beta_1 = \frac{\varepsilon_1 \mu_1 W_1}{L_1 t_{ox1}} \tag{4-2}$$

$$V_1 = V_{G1S} - V_{P1} \tag{4-3}$$

on the condition of

$$V_1 \geqq V_M \tag{4-4}$$

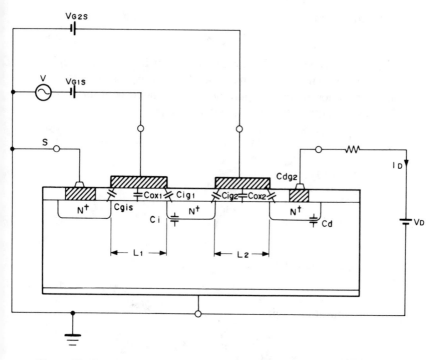

Figure 53. Cross section of a dual-gate MOSFET with capacitive elements designated. The biasing circuitry is also shown.

where L_1 is the length of the gate; V_{P1} is the pinch-off voltage; W_1 is the channel width; t_{ox1} and ε_1 are the thickness and the dielectric constant of the gate oxide, respectively; μ_1 is the electron mobility of the channel; V_M is the potential of the island, I_{D1} is the channel current; and V_{GIS} is the applied bias between the first gate and the source.

In the saturation region, the drain current (I_{DS1}) is expressed as

$$I_{DS1} = \frac{\varepsilon_1 \mu_1 W_1}{2 L_1 t_{ox1}} (V_{GIS} - V_{P1})^2$$
$$= \frac{\beta_1}{2} V_1^2 \tag{4-5}$$

on the condition of

$$V_1 \lesseqgtr V_M \tag{4-6}$$

As for the second FET, the drain current in the triode region (I_{D2}) and that in the saturation region (I_{DS2}) are expressed as follows (subscript 2 designates the parameters associated with the second FET and V_D designates the drain bias voltage):

$$
\begin{aligned}
I_{D2} &= \frac{\varepsilon_2 \mu_2 W_2}{2 L_2 t_{ox2}} [(V_{G2S} - V_{P2} - V_M)^2 - (V_{G2S} - V_{P2} - V_M - V_D \\
&\quad + V_M)^2] \\
&= \frac{\beta_2}{2} [(V_2 - V_M)^2 - (V_2 - V_D)^2 \tag{4-7}
\end{aligned}
$$

$$\beta_2 = \frac{\varepsilon_2 \mu_2 W_2}{L_2 t_{ox2}} \tag{4-8}$$

$$V_2 = V_{G2S} - V_{P2} \tag{4-9}$$

on the condition of

$$V_2 \gtreqless V_D \tag{4-10}$$

In the saturation region,

$$
\begin{aligned}
I_{DS2} &= \frac{\varepsilon_2 \mu_2 W_2}{2 L_2 t_{ox2}} (V_{G2S} - V_{P2} - V_M)^2 \\
&= \frac{\beta_2}{2} (V_2 - V_M)^2 \tag{4-11}
\end{aligned}
$$

on the condition of

$$V_M \lesseqgtr V_2 \lesseqgtr V_D \tag{4-12}$$

The operation is divided into four regions according to whether the first or the second gate is in the triode mode or in saturation. The operation modes are summarized in Table 13.

Table 13. The operation modes of a dual-gate MOSFET

Region	First FET	Second FET	Operation mode
I	Triode	Tride	
II	Triode	Saturation	Gain reduction
III	Saturation	Triode	
IV	Saturation	Saturation	Full Gain

At full gain, both the first and the second FETs are in saturation. As the gain is reduced by applying a bias toward the cutoff of the second gate, the first FET operates in the triode region.

E. F. McKeon reported that the cross-modulation characteristics are improved by applying a bias to the first gate toward the positive direction while the second gate is approaching cutoff. In this way, the first FET is operated into the deeper triode region, that is, it is more forward biased.

4.2.2. Cross Modulation in the Gain-Control Mode

4.2.2.1. Cross Modulation as a Function of Transconductance

The small signal current (i_d) is expressed as a function of the transconductance (g_{m1}) and the small signal voltage (v_g) applied to the first gate:

$$i_d = g_{m1}v_g + \frac{\partial g_{m1}}{\partial V_{G1S}} v_g^2 + \frac{\partial^2 g_{m1}}{\partial V_{G1S}^2} v_g^3 + \cdots \tag{4-13}$$

The cross-modulation distortion (m_K) is mainly caused by the second derivative of g_m and is expressed as

$$m_K = \frac{v_g^2}{2g_{m1}} \frac{\partial^2 g_{m1}}{\partial V_{G1S}^2} \tag{4-14}$$

Therefore, in order to investigate the cross-modulation distortion in the gain-control operation, the second derivative of the transconductance to the first gate should be analyzed.

4.2.2.2. Correspondence of Cross Modulation and g_m Curves

As an example, the correspondence of cross-modulation distortion and the transconductance curves was investigated for a dual-gate MOSFET (3SK48). The typical structural parameters and characteristics are summarized in Table 14.

The transconductance curves as functions of the first and second gate bias voltages are shown in Figure 54, and the corresponding cross-modulation characteristics in Figure 55.

The cross-modulation distortion is usually evaluated by the am-

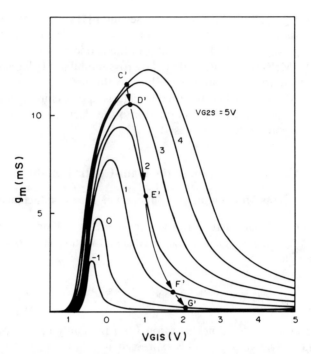

Figure 54. Transconductance of a dual-gate MOSFET (3SK48) as a function of the first gate bias voltage with the second gate bias voltage as a parameter. C′, D′, E′, F′, and G′ represent the locus of the bias points when automatic gain control is applied.

Table 14. Structural parameters and characteristics of the 3SK48

Parameter	Minimum	Typical	Maximum
First gate length (μm)		2.4	
Second gate length (μm)		3.0	
Gate width (μm)		800	
Input capacitance (pF)		1.8	2.5
Feedback capacitance (fF)		10	20
I_{DSS} (mA)	4		11
g_m (mS)	9	11	
V_{P1} (V)	−3.0		
V_{P2} (V)	−3.0		
Power gain (dB at 800 MHz)	8	13	
Noise figure (dB at 800 MHz)		3.7	5.0

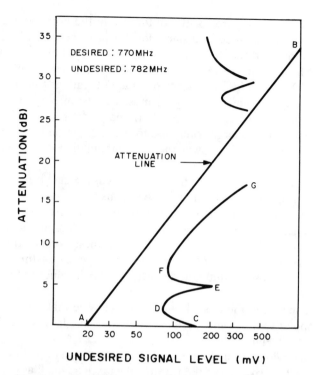

Figure 55. Cross-modulation characteristics of a dual-gate MOSFET (3SK48). Points C, D, E, F, and G correspond to the bias points C', D', E', F', and G' in Figure 54.

plitude of an interfering (undesired) signal that causes 1% of cross modulation to the desired signal. The frequency of the undesired signal is 12 MHz away from the desired, which in practice corresponds to the adjacent channel.

The level of the undesired signal is plotted against the attenuation of the desired signal in Figure 55. In the AGC (automatic gain-control) mode of tuners, gain reduction is applied through the gate biases so that the output level of the tuner stays the same even if the input level exceeds the AGC starting point (A). In Figure 55, the point is at 20 mV. The gain control is applied to the desired signal, and thus the desired signal level is expressed by straight line AB in the figure with a slope of 6 dB per octave.

Therefore, in Figure 55, the difference in the horizontal direction between the undesired signal level, which causes 1% of cross modulation, and the desired signal level expressed by the straight line (AB) is the measure of the cross-modulation distortion, and is called the D/U ratio. In other words, the D/U ratio is a measure of how strong the undesired signal must be to cause 1% of cross modulation to the desired signal. If the undesired level in the figure in relatively large compared to the attenuation line (AB) of the desired signal, the lower the cross-modulation distortion will be.

In Figure 54, the locus of the bias points as the gain is reduced is plotted as C′, D′, E′, F′, and G′. At points C′, E′, and G′, the g_m curve is straight and has no curvature, giving a zero value to the second derivative of g_m. This suggests that the cross modulation is very small or the D/U ratio is very large at these points. At points D and F, the curvature of the g_m curve is large, suggesting a relatively large cross-modulation distortion or small values of the D/U ratio.

By comparing Figures 54 and 55, we can see that these points correspond to the points designated by the same letters, C, D, E, F, and G in Figure 54.

As is obvious from Equation (4-14), a structural design to make the second derivative of the transconductance as small as possible is required in order to keep the cross modulation at a low level in the AGC mode of operation.

4.2.2.3. Transconductance Curve and Structural Parameters

The ratio of the current-driving parameters (β_2/β_1) is taken as a parameter in order to analyze the second derivative of the transconductance, and hence the cross modulation. The value of β reflects the structural parameters, as is seen in Equations (4-2) and (4-8).

R. M. Barsan reported the numerically calculated and measured g_m curves for dual-gate MOSFETs, from which we can see that the curvature of the transconductance curve is small for a large ratio of β_2 to β_1.[23]

Usually, dual-gate MOSFETs are designed so that β_2 can be much smaller than β_1. For example, L_2 was much larger than L_1 in early devices in order to reduce the feedback capacitance by the large voltage-amplification factor of the second FET.

The second derivative of g_m is analyzed in the following to obtain design principles for a low-distortion dual-gate FET.

When the first FET is in saturation, the following equations are derived:

$$g_{mS1} = \frac{\partial I_{DS1}}{\partial V_{G1S}}$$
$$= \beta_1 V_1 \tag{4-15}$$

$$\frac{\partial g_{mS1}}{\partial V_{G1S}} = \beta_1 \tag{4-16}$$

$$\frac{\partial^2 g_{mS1}}{\partial V_{G1S}^2} = 0 \tag{4-17}$$

In the saturation region, the second derivative is zero, giving no third-order distortion. However, if the first gate is biased in the direction of cutoff for gain reduction, it is well known that the cutoff itself gives a large distortion, which is not tolerable in practice.

When the first FET is just at pinch-off ($V_1 = V_M$), the following equations are derived:

$$\frac{\beta_1}{2} V_1^2 = \frac{\beta_2}{2} (V_2 - V_1)^2 \tag{4-18}$$

and, hence,

$$V_1 = \frac{1}{1 + 1/\sqrt{m}} V_2 = V_C \tag{4-19}$$

where

$$m = \beta_2/\beta_1 \tag{4-20}$$

The bias of the first gate, which is just at saturation, is designated as V_C.

In case the first FET is in the triode region and the second is in saturation, which corresponds to the gain-reduced mode, Equations (4-1) and (4-11) are set equal, yielding

$$m(V_G - V_M)^2 = V_1^2 - (V_1 - V_M)^2 \tag{4-21}$$

Equation (4-21) is solved for V_M, which is substituted into Equation (4-11), yielding the equation for drain current I_D:

$$I_D = \frac{\beta_1 m}{2(1 + m)^2} \left[(1 - m)(V_1 - V_2)^2 + (1 + m)V_1^2 \right.$$

$$\left. - 2(V_1 - V_2) \sqrt{(1 + m)V_1^2 - m(V_1 - V_2)^2} \right] \quad (4\text{-}22)$$

Differentiating by V_1, we obtain g_m:

$$g_{m1} = \frac{\beta_1 m}{(1 + m)^2} \left[2V_1 - (1 - m)V_2 - \sqrt{V_1^2 + 2mV_1V_2 - mV_2^2} \right.$$

$$\left. - \frac{(V_1 - V_2)(V_1 + mV_2)}{\sqrt{V_1^2 + 2mV_1V_2 - mV_2^2}} \right] \quad (4\text{-}23)$$

The derivative of g_m is obtained as

$$\frac{\partial g_{m1}}{\partial V_1} = \frac{\beta_1 m}{(1 + m)^2} \left\{ 2 - \frac{2(V_1 + mV_2)}{\sqrt{(1 + m)V_1^2 - m(V_1 - V_2)^2}} \right.$$

$$\left. - \frac{V_1 - V_2}{\sqrt{(1 + m)V_1^2 - m(V_1 - V_2)^2}} \left[1 - \frac{(V_1 + mV_2)^2}{(1 + m)V_1^2 - m(V_1 - V_2)^2} \right] \right\} \quad (4\text{-}24)$$

Then the second derivative of g_m is expressed as

$$\frac{\partial^2 g_{m1}}{\partial V_1^2} = \frac{\beta_1 m}{(1 + m)^2} \left[-\frac{3}{\sqrt{M}} + \frac{3(V_1 + mV_2)^2}{M\sqrt{M}} + \frac{3(V_1 - V_2)(V_1 + mV_2)}{M\sqrt{M}} \right.$$

$$\left. - \frac{3(V_1 - V_2)(V_1 + mV_2)^3}{M\sqrt{M}} \right] \quad (4\text{-}25)$$

where

$$M = (1 + m)V_1^2 - m(V_1 - V_2)^2 \quad (4\text{-}26)$$

If the second gate bias (V_2) is fixed, the operating point of the first FET moves from the triode region to the saturation region at the point where the first gate bias is equal to V_C given by Equation

(4-19) as the first gate bias (V_1) is increased. Transconductance g_m is proportional to V_1, as in Equation (4-15), when V_1 is less than V_C. When V_1 is equal to or larger than V_C, g_m decreases according to Equation (4-23) as the bias is increased. Since transition bias V_C is proportional to V_2, the transconductance curves change in equal steps as the second gate bias V_2 is changed in equal steps, as shown in Figure 55.

Calculated transconductance curves are plotted in Figure 56 with V_C as a parameter. Because V_C is proportional to V_2, parameter V_C is equivalent to V_2. There is a discontinuity of the slope of the g_m curve at the bias equal to V_C ($V_1 = V_C$), and g_m exhibits a sharp apex. In a real device, the corner is smoothed, as shown in the figure by dashed lines.

It can be expected that the slope of the g_m curve at the right side

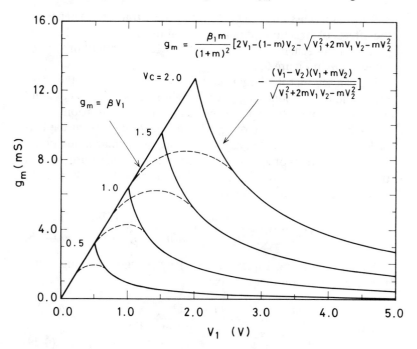

$$g_m = \frac{\beta_1 m}{(1+m)^2}\left[2V_1-(1-m)V_2-\sqrt{V_1^2+2mV_1V_2-mV_2^2}\right.$$
$$\left. - \frac{(V_1-V_2)(V_1+mV_2)}{\sqrt{V_1^2+2mV_1V_2-mV_2^2}}\right]$$

Figure 56. Calculated transconductance curves with V_C as a parameter. It is assumed that $m = 1$ and $\beta = 6.37$ mS/V.

of the apex corresponds to the zero curvature point (E) of Figure 55 with the points of positive and negative curvatures (D and F) at both sides. If the slope is steep or $|\partial g_m/\partial V_1|$ is large, the curvature at both sides also becomes large, giving rise to the second derivative of the transconductance $|\partial^2 g_m/\partial V_1^2|$, and hence the cross-modulation distortion increases.

If V_C is fixed, g_m exhibits the same value at the bias point of V_1 (equal to V_C) regardless of the structural parameter ($m = \beta_2/\beta_1$). Then if the second gate bias V_2 is set as in Equation (4-27), which is equivalent to Equation (4-19), g_m exhibits a peak at the point where V_1 equals V_C regardless of the value of m.

$$V_2 = \left(1 + \frac{1}{\sqrt{m}}\right) V_C \tag{4-27}$$

If the gain is assumed to be determined by transconductance g_m, the effects of structural parameter m on the g_m curve can be compared. In this scheme, g_m in Equation (4-23) is plotted for various values of m in Figure 57.

We can see that the larger the value of m, the smaller the absolute value of the slope of g_m at the right side of the g_m peak, and hence the smaller the expected cross-modulation distortion.

Analytically, the value of $\partial g_m/\partial V_1$ in Equation (4-24) at the bias of V_1 equal to V_C is taken as the measure of the magnitude of the second derivatives of g_m at both sides of the zero curvature point of the g_m curve. The value of the $\partial g_m/\partial V_1$ at the bias of $V_1 = V_C$ and V_2, as in Equation (4-27), is derived as

$$\frac{\partial g_{m1}}{\partial V_1}\bigg|_{V_1=V_C} = -\beta_1 \frac{2m^{5/2} + m^2 + 4m^{3/2} + 2m + 2m^{1/2} + 1}{m(m + 1)^2} \tag{4-28}$$

By differentiating Equation (4-29) by m, the following is obtained on condition that β_1 is fixed:

$$\frac{\partial}{\partial m}\left(\frac{\partial g_{m1}}{\partial V_1}\right) = \beta_1 \frac{m^{7/2} + m^3 + 3m^{5/2} + 3m^2 + 3m^{3/2} + 3m + m^{1/2} + 1}{m^2(m + 1)^3} \tag{4-29}$$

When the value of m is unity, the following is obtained:

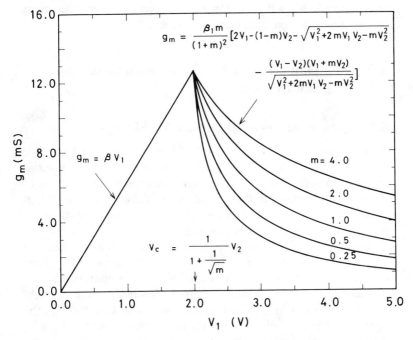

Figure 57. Calculated transconductance curves with m as a parameter. It is assumed that $V_C = 2$ and $\beta = 6.37$ mS/V.

$$\frac{\partial}{\partial m}\left(\frac{\partial g_{m1}}{\partial V_1}\right) = 2\beta_1 > 0 \qquad (4\text{-}30)$$

Therefore, as the value of m increases, the value of $|\,\partial g_{m1}/\partial V_1\,|$ decreases, and, consequently, the cross modulation is improved at the points corresponding to D and F in Figure 55.

We can see that the slope of the g_m curve at the right side of the g_m peak decreases as the value of m increases.

4.2.2.4. Design Principles for Low Cross-Modulation Distortion

It has been shown that increasing m or β_2/β_1 can result in the improvement of cross modulation. Since β_2 is a function of several structural parameters, there could be a number of ways to increase the value of β_2, and, consequently, the value of m. The most common is to design L_2 as small as possible.

4.3. Dual-Gate MOSFET with Grooved Gates and Deep Ion Implantation

4.3.1. Requirements for Dual-Gate MOSFETs

Basic requirements for a dual-gate MOSFET are the low noise figure and the high associated gain at high frequencies. For this purpose, the cutoff frequency, which is determined by the ratio of the transconductance to the capacitance of the first FET (g_{m1}/C_{ox1}), should be as high as possible. Therefore, the channel length of the first gate is made as short as possible.

According to the analysis in the previous section, the channel length of the second gate should be minimized in order to reduce cross-modulation distortion in the AGC mode of operation.

Reduction of the channel length is limited by the "punch-through" effect, which allows a large current to flow through the injection of carriers directly into the substrate toward the drain by the electric field extended from the drain. This effect gives a larger constraint to the second gate because the electric field created by the drain is largest when the second gate approaches pinch-off in the deeply gain-controlled region.

One of the structures effective in reducing the electric field around the drain and thereby achieving the short gate length is the grooved gate structure,[7,24-26] which is shown in Figure 58. In this configuration, the drain and the island regions are placed in the upper level so that the side walls of diffused regions do not face each other, reducing the electric field across the channel. The structure is desirable for both the first and the second gates.

Another means to limit the extension of the drain depletion region is to employ a substrate with a high impurity concentration. However, the substrate effects increase in this case, and, for example, g_m decreases, thus degrading the noise figure.

The effective structure is obtained by the introduction of a high concentration region deep under the second channel.[7] The im-

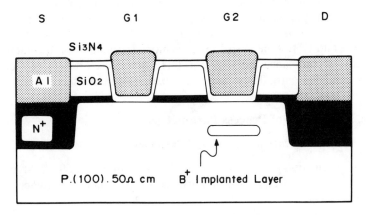

Figure 58. Cross section of a dual-gate MOSFET with grooved gates and deep ion implantation under the second gate.

planted region limits the extension of the electric field from the drain, and at the same time the region of the substrate adjacent to the channel remains at a low concentration. This structure is also shown in Figure 58. The parameters for implantation, such as the energy and dose, become of practical interest. In this section, the most appropriate design of the ion implantation is studied by experiments based on the grooved gate structure.

Another characteristic of interest is the feedback capacitance (C_{dg1}), which should be as small as possible. The reduction of the drain electric field is also effective in decreasing the feedback capacitance because it is expressed by the capacitance between the first gate and the island (C_{ig1}) divided by the voltage amplification factor (μ_2) of the second FET as follows:

$$C_{dg1} = \frac{C_{ig1}}{1 + g_{m2} \cdot r_{d2}}$$
$$\doteqdot \frac{C_{ig1}}{\mu_2} \tag{4-31}$$
$$\mu_2 = g_{m2} \cdot r_{d2} \tag{4-32}$$

where g_{m2} and r_{d2} designate the transconductance and the output resistance of the second FET, respectively, and the product of $g_{m2} \cdot r_{d2}$ is the voltage-amplification factor.

The drain bias, which can fully activate the device, is another important parameter as lower power consumption is steadily required. And if the same dual-gate MOSFET is usable both for VHF and UHF tuners, the device is most desirable from the practical viewpoint since the VHF and UHF tuners are often integrated on a single board. These issues are examined in this section.

4.3.2. Design of Process Parameters

4.3.2.1. Fabrication Process and Parameters

The fabrication process of the FET is shown in Figure 59. A P-type (100) silicon wafer with an impurity concentration of 3×10^{14} cm^{-3}

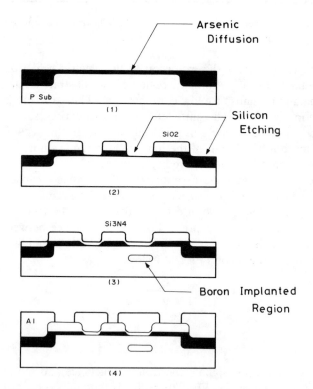

Figure 59. Fabrication process of the dual-gate MOSFET.

was employed as the substrate. The PNP back-to-back diodes for gate protection were formed on the wafer prior to the process for the FET.

The first step was to form relatively deep N-type diffused regions for ohmic contacts of the source and drain, followed by a shallow diffusion of arsenic into the whole surface of the wafer. Then the SiO_2 of the channel regions as well as that of the source and drain ohmic contact areas and the rest of the silicon surface were etched off, followed by plasma etching of the silicon surface doped with arsenic. The channel lengths of the first and the second FETs were 2.3 and 2.8 microns, respectively.

Deep ion implantation of boron was performed through the photoresist film as a mask into the deep regions of silicon under the second channel. Then the gate oxide is formed by thermal oxidation to the thickness of 400 Å, followed by a deposition of silicon nitride of 500 Å to assure stability against mobile ions.

4.3.2.2. g_{m2} and r_{d2} in Relation to Ion-Implantation Conditions

In Figure 60, the relation of g_{m2} and r_{d2} is plotted with ion-implantation conditions as parameters. The lines exhibit the loci of the constant voltage-amplification factor (μ_2) of the second FET. Deep ion implantation has a remarkable effect in increasing the output resistance with a slight degradation of transconductance, thereby resulting in a great improvement of the voltage-amplification factor, and, consequently, reducing the feedback capacitance.

Shallow implantation, by contrast, has only a marginal effect.

In Figure 61, feedback capacitance C_{dg1} is plotted as a function of the second gate bias voltage V_{G2S} with dose as a parameter. The implantation energy was fixed at 360 keV.

As the dose increases, the feedback capacitance decreases, reflecting the increase of the output impedance of the second FET. With doses of 2×10^{11} cm^{-2} and 3×10^{11} cm^{-2}, the capacitance increases sharply for the second gate bias of 2.5 to 3 volts. These values of bias voltages correspond to the potential of the island (approximately expressed as $V_{G2S} - V_{P2}$) at which the depletion layer spreads beyond the implanted region with relatively high concentration. The biases correspond well to the calculated values.

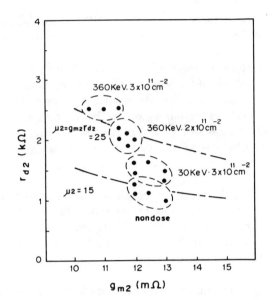

Figure 60. Relation between g_{m2} and r_{d2} with implantation energy and dose as parameters.

In Figure 62, the same characteristics are plotted with ion implantation energy as a parameter. The dose was fixed at 3×10^{11} cm^{-2}. The feedback capacitance again decreases as the energy is increased, reflecting the increase of output impedance. The sharp rise of the capacitance with an increase of the bias beyond a certain point corresponds to the spread of the depletion layer beyond the implanted region.

If the dose is too high, the transconductance decreases, affecting other parameters such as noise figure. If the implantation energy is too high, the effect of suppressing the extension of the depletion layer becomes small. An energy of 360 keV and a dose of 2.5×10^{11} cm^{-2} were chosen as the optimum ion implantation conditions.

In Figure 63, the feedback capacitance is plotted against the drain bias voltage. At a bias of 6 volts, the capacitance was 22 fF, reduced by 10 fF as a result of deep ion implantation.

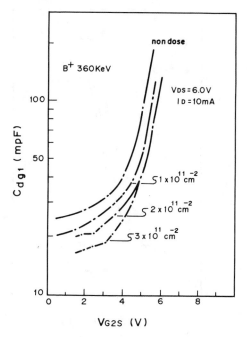

Figure 61. Feedback capacitance C_{dg1} as a function of the second gate bias voltage V_{G2S} with ion dose as a parameter.

4.3.2.3. The Package

The TO-72 with a metal base was employed as the package. The thickness of the metal base was made as thin as 1 mm to minimize the parasitic inductance of the outer leads.

4.3.2.4. The Parameters

The structural parameters and associated electrical characteristics are summarized in Table 15.

4.3.3. Device Characteristics

The drain current is plotted as a function of drain bias with the first gate bias as a parameter in Figure 64. The transconductance curve

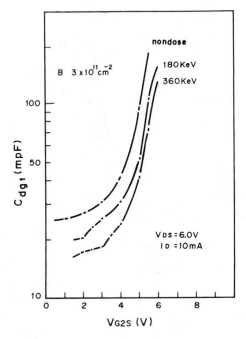

Figure 62. Feedback capacitance C_{dg1} as a function of the second gate bias voltage V_{G2S} with implantation energy as a parameter.

Table 15. Structural parameters and characteristics

Parameter	Minimum	Typical	Maximum
First gate length (μm)		2.3	
Second gate length (μm)		2.8	
Gate width (mm)		1.36	
Input capacitance (pF)		4.5	
Feedback capacitance (fF)		20	
I_{DSS} (mA)	4		11
g_m (mS)	9	11	
V_{P1} (V)			−3.0
V_{P2} (V)			−3.0

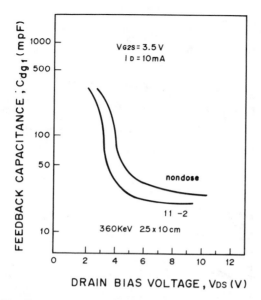

Figure 63. Feedback capacitance C_{dg1} as a function of drain bias voltage V_{DS}. The implantation energy is 360 keV and the dose is $2.5 \times 10^{12} \times$ cm^{-2}.

is shown in Figure 65. The locus of the bias point in the AGC mode is plotted in the figure with small circles.

The values of the power gain and noise figure at 200 and 800 MHz are summarized in Table 16.

4.3.4. Cross-Modulation Characteristics

Cross-modulation characteristics were measured at 200 and 800 MHz with the undesired frequencies of 212 and 812 MHz, respectively. The results are plotted in Figure 66. A bias circuit that allows the first gate bias to move in the forward direction as the second gate bias is applied toward cutoff is also shown in the figure. We can see that the cross-modulation characteristics are very good in the small-to-moderate gain-reduction range, where the second derivative of the transconductance plays a major role in generating third-order distortion.

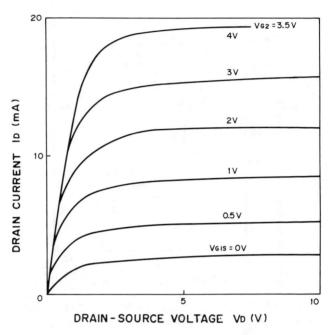

Figure 64. Drain-current characteristics with the first gate bias voltage as a parameter.

Table 16. Power gain and noise figure at VHF and UHF

Parameter	Value at 200 MHz	Value at 800 MHz
Power gain (dB)	22.0	12.5
Noise figure (dB)	1.4	3.5

V_D = 6.0 V, I_D = 10mA, and V_{G2S} = 3.5 V.

4.3.5. Summary of the Characteristics

A novel structure of a dual-gate MOSFET with grooved gates and deep ion implantation under the second channel was investigated. The design was effective in reducing the second gate length of a dual-gate MOSFET and thus theoretically in decreasing the cross-modulation distortion under gain-controlled operation. It was shown that the output impedance of the second FET becomes

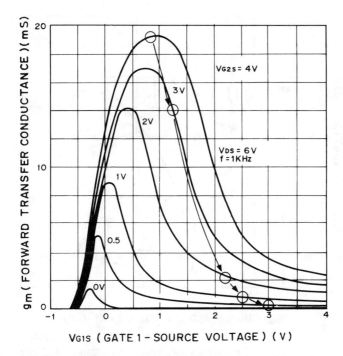

Figure 65. Transconductance curve with the second gate bias voltage as a parameter. The locus of the bias points in the automatic gain-control mode is plotted.

sufficiently large with a very slight sacrifice of transconductance, resulting in a small feedback capacitance. The power gain and noise figure exhibited good values, and the cross-modulation characteristics were also very good at both the VHF and UHF bands.

4.4. Dual-Gate GaAs MESFET with Low Distortion

4.4.1. Requirements for Low Noise and Low Distortion

As described in the previous section, the silicon MOSFET has limitations in noise figure at UHF frequencies. On the other hand, the requirement for low-noise characteristics is stronger in the

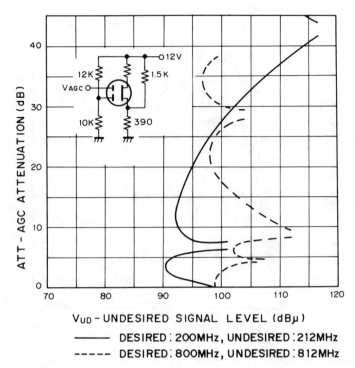

Figure 66. Cross-modulation characteristics measured at 200 and 800 MHz. The bias circuit is also shown.

UHF range as the propagation of the electromagnetic wave is often hindered by objects along the path, resulting in a much weaker signal strength at the reception site.

In order to meet this requirement, the dual-gate GaAs MESFET has been extensively studied.[18–22] An example of the overall noise-figure characteristics of a tuner covering the VHF and UHF bands is shown in Figure 67. The VHF band was covered by a dual-gate MOSFET and the noise figure in the UHF band was measured for both a dual-gate MOSFET and a dual-gate GaAs MESFET (3SK147). We can see that the dual-gate GaAs MESFET improves the noise figure in the UHF band to the same level as in the VHF band.

However, the device suffers from inferior cross-modulation char-

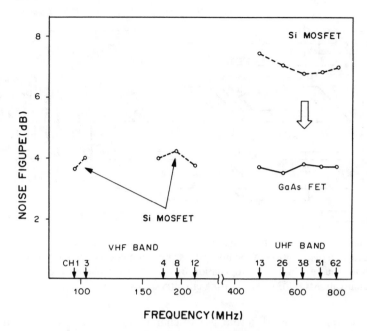

Figure 67. Noise figure of an electronic tuner covering the VHF and UHF bands. Comparison is made between the silicon dual-gate MOSFET and the dual-gate GaAs MESFET when applied to the first-stage amplifier of the tuner.

acteristics compared with its silicon counterpart. As was discussed in Section 4.2, the cross modulation in the gain-controlled region can be improved by appropriately designing the structure of the gates.

One of the most realizable designs is to set the ratio of the channel lengths ($m = L_1/L_2$) as large as possible. In the design of a MOSFET, the length of the first gate has to be minimized to achieve a low noise figure. Therefore, the goal is to reduce the second channel length as much as possible.

In designing a GaAs FET, in contrast, the length of the first channel can be made large in comparison with that of the second because electron mobility is high enough that the noise figure has some tolerance.

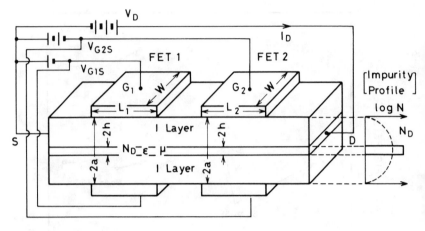

Figure 68. Schematic diagram of a dual-gate GaAs MESFET. The gaussian distribution of the impurity is approximated by a narrow distribution, thus enabling the application of the same calculation used for the dual-gate MOSFET. The biasing circuitry is also shown.

In this section, different sets of channel lengths are investigated in order to develop a structure that exhibits both a sufficiently low noise figure and cross-modulation characteristics.

4.4.2. Design[22]

The schematic diagram of a dual-gate MESFET is shown in Figure 68 with structural and biasing parameters. If the impurity profile formed by ion implantation is modeled as a rectangular region with very narrow widths, as in the figure, the characteristics become similar to that of a MOSFET and the following equations are derived in accordance with the analysis of Section 4.2:

$$g_m = \beta_1 V_1 \qquad (V_1 < V_C) \tag{4-33}$$

$$g_m = \frac{\beta_1 m}{(1+m)^2} \left[2V_1 - (1-m)V_2 - \sqrt{V_1^2 + 2mV_1V_2 - mV_2^2} \right.$$
$$\left. - \frac{(V_1 - V_2)(V_1 + mV_2)}{\sqrt{V_1^2 + 2mV_1V_2 - mV_2^2}} \right] \qquad (V_1 \geqq V_C) \tag{4-34}$$

where

$$V_C = V_2/(1 + 1/\sqrt{m}) \qquad (4\text{-}35)$$
$$m = \beta_2/\beta_1 \qquad (4\text{-}36)$$
$$\beta = \varepsilon \mu W/aL \qquad (4\text{-}37)$$
$$V_{1,2} = V_{G1,2} - V_{P1,2} \qquad (4\text{-}38)$$

and subscripts 1 and 2 correspond to the first and second gate, respectively.

The behavior of the transconductance is shown in Figure 69. As V_1 is increased, g_m takes its maximum value at $V_1 = V_C$ and then decreases in the region $V_1 \gtrsim V_C$. As V_2 decreases and G_2 approaches pinch-off, the g_m curve shrinks proportionately. Actual curves are rounded off at the tops, as indicated in the figure by dashed lines. As m decreases, the negative slope of g_m at V_1 slightly

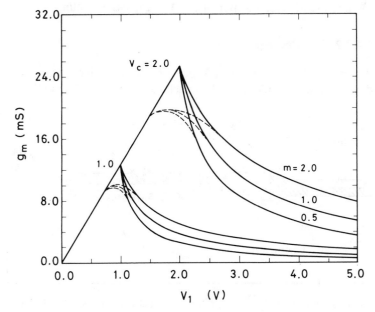

Figure 69. Calculated transconductance curves with V_C and m as parameters. It is assumed that β is equal to 13.7 mS/V.

larger than V_C becomes steeper, corresponding to the rise of the second derivatives of g_m at both sides of the zero-curvature point.

Therefore, one can adopt the design principle to set m ($= \beta_2/\beta_1$) as large as possible to reduce cross-modulation distortion in the small-to-moderate gain-control region.

4.4.3. Device Fabrication

4.4.3.1. Structure and Fabrication Process

The cross-sectional structure of the device is shown in Figure 70. For protection of the gates, back-to-back diodes are fabricated on the chip with a P+N+P+ structure.

The fabrication process is shown in Figure 71. Protection diodes were formed prior to the formation of the active region. The channel was formed by ion implantation of silicon. The annealing of the ion-implanted layer was performed by "capless annealing."[27]

4.4.3.2. The Package

A very small plastic-mold package with four outer leads was newly developed and employed.[19] The package is shown in Figure 72.

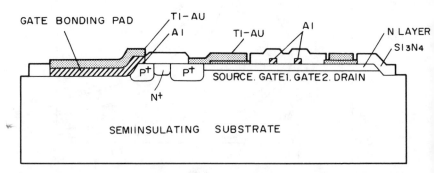

Figure 70. Cross section of the fabricated dual-gate GaAs MESFET with on-chip protection diodes.

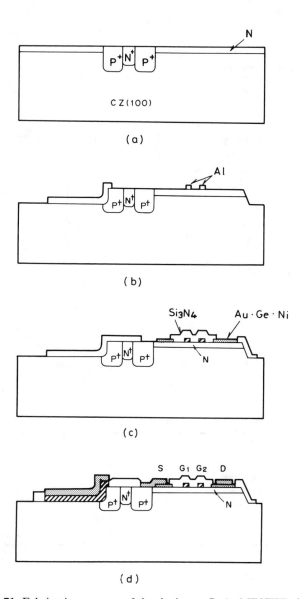

Figure 71. Fabrication process of the dual-gate GaAs MESFET. (a) Protection diodes and active layer formation. (b) Mesa etching of the active layer and aluminum gates formation. (c) Ohmic alloy formation and passivation with Si_3N_4 film. (d) Bonding pads formation.

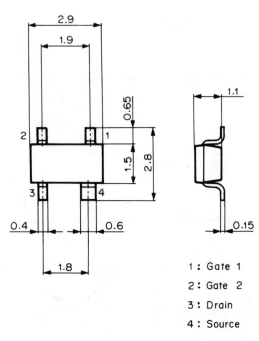

1 : Gate 1
2 : Gate 2
3 : Drain
4 : Source

Figure 72. Plastic-mold miniature-sized package with four outer leads.

Table 17. Structural parameters and characteristics

Parameter	Model 4a	Model 4b	Model 4c	Model 4d
Channel width (μm)	400	400	400	400
Source–drain distance (μm)	15	15	15	15
First gate length (μm)	3.2	2.2	1.2	1.2
Second gate length (μm)	1.2	1.2	1.2	2.2
$m \ (= \beta_1/\beta_2)$	2.67	1.83	1.0	0.56
V_P (V)	-1.8	-1.8	-1.8	-1.8

4.4.3.3. Fabricated Devices

Four devices were fabricated with different sets of channel lengths, as shown in Table 17. The gate lengths shown in the table are measured values. The source-to-drain distance is 15 microns for all the devices.

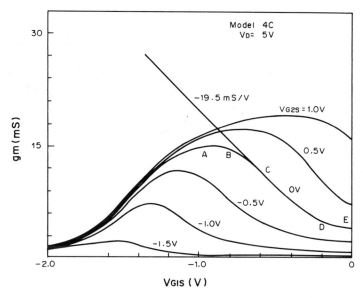

Figure 73. Measured transconductance curve of device model 4c with the second gate bias as a parameter. The maximum value of the negative slope on the curve having the highest g_m of 15 mS is -19.5 mS/V.

4.4.4. Cross-Modulation Characteristics

As an example, the g_m curve of FET 4c is shown in Figure 73, where the corresponding maximum value of the negative slope of g_m is shown. The values of the negative slopes are plotted for devices with different values of m in Figure 74. As the value of m increases, the value of the negative slope decreases, resulting in a more gradual change of g_m and, consequently, the decrease of the second derivatives of g_m at both sides of the zero-curvature point, which is expected from the analysis.

The cross-modulation characteristic of the same device is shown in Figure 75. Points A to E in Figure 73 correspond to the same operating conditions as points A' to E' in Figure 75.

The amount of gain control corresponds to the strength of the incoming desired signal. In Figure 75, the strength of the desired

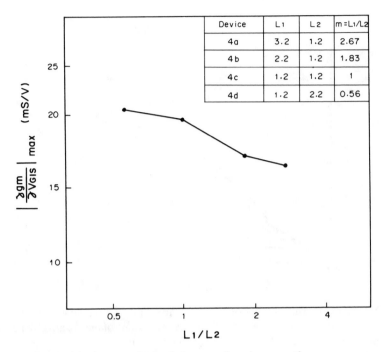

Figure 74. Maximum values of the negative slope on the g_m curve are plotted against the ratio L_1/L_2 ($=m$).

signal is divided into three regions. The weak-signal region corresponds to a gain reduction to 0 to 10 dB, the medium to that of 10 to 20 dB, and the strong to that of 20 to 30 dB. The level of the undesired signal, which causes a cross-modulation distortion of 1%, is plotted in the figure.

The level of the desired signal is expressed by a straight line, with the AGC starting at −50 dBm. In Figure 75, the ratio of the undesired signal level to the desired is referred to as the D/U ratio, and reflects the amount of cross modulation. In the figure, the values of the minimum, or the worst, D/U ratio in the weak-, medium-, and strong-signal regions are shown as D/U 1, D/U 2 and D/U 3, respectively.

In Figure 76, the D/U ratios are plotted as a function of the

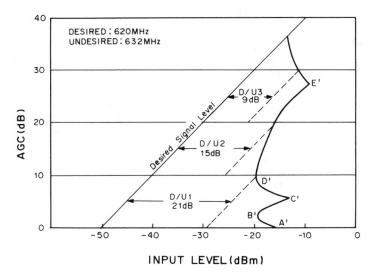

Figure 75. Cross-modulation characteristics of dual-gate MESFET model 4c. The bias points designated as A to E in Figure 73 correspond to points A' to E' in this figure.

value of m. In the weak- and medium-signal regions, the ratios increase with increasing m, supporting the design principle. However, D/U 3 shows the opposite result.

In the region where the signal is strong, the origin of the cross modulation is expected to lie in the cutoff characteristics of the second FET because the large gain reduction is achieved mainly by the second gate, which is completely cut off. However, the incoming signal is applied to the first gate and is attenuated by the first FET, which is operating in the triode region.

The FET in the triode region can be regarded as a distributed line consisting of a distributed capacitance and resistance. Therefore, the amount of attenuation can be roughly estimated by the gate capacitance and the channel resistance, both of which are proportional to channel length.

In Figure 77, the values of D/U 3 are plotted against the first channel length. Note that the values of D/U 3 are the same be-

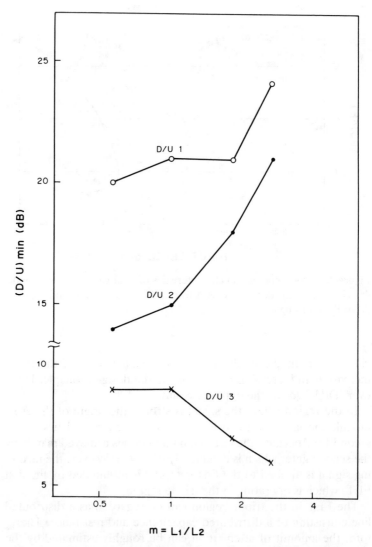

Figure 76. Values of the worst D/U ratios at the three levels of attenuation range are plotted against the value of m. The values of D/U 1 correspond to the attenuation range of 0 to 10 dB, those of D/U 2 to that of 10 to 20 dB, and those of D/U 3 to 20 to 30 dB.

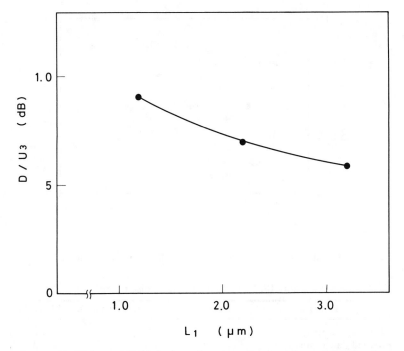

Figure 77. Values of D/U 3 plotted against the first channel length, L_1.

tween devices 4c and 4d, which have the same first-channel length. We can see that as the length of the first gate (L_1) increases, the value of D/U 3 decreases, suggesting a corresponding decrease of the attenuation by the first FET.

We can also presume that D/U 3 is affected by the cutoff characteristics of the second gate, such as V_P, and the remoteness of the cutoff. In Figure 78, D/U 3 is shown as a function of V_P, where we can see that increasing V_P improves D/U 3. In this way, D/U 3 can be made large and independent of m.

4.4.5. Power Gain and the Noise Figure

Power gain and the noise figure at 800 MHz were measured and listed in Table 18. Since GaAs MESFETs have a sufficiently low

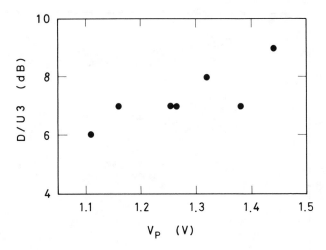

Figure 78. Values of D/U 3 plotted against the pinch-off voltage, V_P.

Table 18. Power gain and the noise figure[a]

Parameter	Model 4a	Model 4b	Model 4c	Model 4d
First gate length (μm)	3.2	2.2	1.2	1.2
Second gate length (μm)	1.2	1.2	1.2	2.2
$m\ (= \beta_1/\beta_2)$	2.67	1.83	1.0	0.56
Power gain (dB at 800 MHz)	17.3	18.0	18.9	19.1
Noise figure (dB at 800 MHz)	1.77	1.53	1.39	1.43

[a]Power gain and the noise figure were measured at biases of $V_D = 5$ V and $I_D = 10$ mA.

noise figure at UHF, the design principle of setting L_1, the length of the first channel, larger than L_2, that of the second, in order to decrease cross-modulation distortion is effective and justifiable.

4.5. Summary

The third-order distortion of the dual-gate FET in the gain-controlled operation mode was analyzed with respect to the structural parameters in order to derive a design principle for low-

distortion devices. Both the dual-gate MOSFET and the dual-gate GaAs MESFET were fabricated with structures suggested as suitable for low noise and low distortion by the design principle; they have given good performance.
The following conclusions were obtained.

1. The magnitude of the second derivative of the transconductance (g_m) of the dual gate FET depends on the ratio (m) of the current-driving parameter of the second FET (β_2) and that of the first (β_1), both of which are determined by structural parameters. The larger the ratio, the smaller the second derivative of g_m, thus resulting in smaller cross-modulation distortion in the small-to-moderate gain-reduction range.

2. In practice, it is recommended that the ratio of the channel lengths $(m = L_1/L_2)$ be as large as possible, or that the channel length of the second gate be as small as possible.

3. In order to achieve a short channel length for the second gate, a grooved gate structure with deep ion implantation under the second gate was employed for the dual-gate MOSFET. This structure is effective in reducing the feedback capacitance and in improving the high-frequency performance both at VHF and UHF.

4. Various sets of the first and the second channel lengths were investigated for the dual-gate GaAs MESFET. The optimum combination for satisfying the requirements of both a low noise figure and low distortion lies in the first-channel length slightly larger than the second, because the GaAs MESFET has a sufficiently low noise figure at the UHF band.

References

1. D. M. Griswold. Characteristics and Applications of RCA Insulated-Gate Field Effect Transistors. *IEEE Trans. Broadcast and TV Receivers* (July 1965): 9–17.
2. R. Dawson, R. Ahrons, and N. Ditrick. Understanding and Using the Dual Gate MOSFET. *Electron. Engrg.* **35** (1967).
3. E. F. McKeon. Cross-Modulation Effects in Single Gate and Dual Gate Field-Effect Transistors. RCA Application Note AN3435 (1967).

4. S. Watanabe. Japanese Patent 51-44067 (1967).
5. S. Watanabe. Japanese Patent 52-28548 (1977).
6. S. Watanabe. Japanese Patent 56-39062 (1981).
7. A. Kayanuma, K. Suzuki, Y. Tsuda, S. Watanabe, and S. Saiki. Low Voltage Operation Dual Gate MOSFETs with Deep Ion Implantation, in *Extended Abstracts of the 15th Symposium on Semiconductors and Integrated Circuits Technology*, pp. 138–143, The Electrochemical Society (1978).
8. T. Okabe, S. Ochi, and H. Kurono. Short Channel, Low Noise UHF MOSFET's Utilizing Molybdenum-Gate Masked Ion-Implantation, in *Proceedings of the 7th Conference on Solid State Devices*, pp. 201–206, Tokyo, 1975. Supplement to *Jpn J. Appl. Phys.* **15** (1976).
9. F. M. Klaassen, H. J. Wilting, and W. C. J. de Groot. A UHF MOS Tetrode with Polysilicon Gate. *Solid State Electron.* **23** (1980): 23–30.
10. T. Okabe and S. Ochi. A Design Consideration of High Frequency Low Noise MOSFET's. *Trans. IECE* **C-39** (1980): 282–289.
11. M. Miyano and S. Umebachi. Low-Noise UHF MOSFET 3SK118 with New Gate Structure. *Nat. Tech. Rep.* **29**, 2(1983): 136–142.
12. M. Miyano, S. Hashizume, and S. Umebachi. High-Frequency MOSFET Series. *Natl. Tech. Rep.* **32**, 2(1986): 11–17.
13. Y. Nishimura, Y. Ogawa, M. Ishino, and S. Matsuura. Super Low Noise Dual Gates MOS FET for TV Tuner. *NEC Gihou* **39**, 3(1986): 82–84.
14. C. A. Liechti, E. Gowen, and J. Cohen. GaAs Microwave Schottky-Gate Field-Effect Transistor, in *Digest of Technical Papers*, pp. 158–159, 1972 International Solid State Circuit Conference (1972).
15. S. Asai, H. Kurono, S. Takahashi, M. Hirao, and H. Kodera. Single- and Dual-Gate GaAs Schottky-Barrier FETs for Microwave Frequencies. Supplement to *Jpn J. Appl. Phys.* **43** (1974): 442–447.
16. C. A. Liechtie. Performance of Dual-Gate GaAs MESFET's as Gain-Controlled Low-Noise Amplifiers and High-Speed Modulators. *IEEE Trans. Microwave Theory Tech.* **MTT-23**, 6 (1975): 461–469.
17. M. Ogawa, K. Ohata, T. Furutsuka, and N. Kawamura. Submicron Single-Gate and Dual-Gate GaAs MESFET's with Improved Low Noise and High Gain Performance. *IEEE Trans. Microwave Theory Tech.* **MTT-24**, 6 (1976): 300–305.
18. T. Sato, A. Hashima, S. Nanbu, G. Kanô, H. Takaoka, and T. Nishiguchi. Low Noise UHF Varactor Tuner with Dual-Gate GaAs MESFET. *Natl. Tech. Rep.* **26**, 2(1980): 334–342.
19. K. Suzuki, M. Kanazawa, T. Aoki, and S. Watanabe. Discrete Plastic Mold FET. *Denshigijutu* **26**, 13(1984): 63–66.

20. Y. Miyawaki, T. Asano, M. Fujita, and I. Shimada. Ion-Implanted Low Noise Deal-Gate GaAs MESFET. *Sanyo Tech. Rev.* **18,** 2(1986): 76–84.
21. H. Ishiuchi, Y. Yamaguchi, T. Sugiki, T. Matsumoto, and K. Kumagai. Dual Gate GaAs FET for a UHF Tuner. *NEC Gihou* **40,** 5(1987): 40–42.
22. S. Watanabe, S. Tanaka, J. Kobayashi, H. Ohke, H. Takakuwa, and O. Yoneyama, Dual Gate GaAs MESFET with Low Distortion Gain Control. *Trans. IEICE* **E 72,** 4(1989): 310–312.
23. R. M. Barsan. Analysis and Modeling of Dual-Gate MOSFET's. *IEEE Trans. Electron Dev.* **ED-28,** 5(1976): 523–533.
24. S. Watanabe. Japanese Patent 47-34096 (1972).
25. S. Watanabe. Japanese Patent 48-36987 (1973).
26. S. Nishimatsu, Y. Kawamoto, H. Masuda, R. Hori, and O. Minato. Grooved Gate MOSFET, in *Proceedings of the 8th Conference on Solid State Devices,* Tokyo, 1976. Supplement to *Jpn J. Appl. Phys.* **16** (1977): 179–183.

CHAPTER 5

Mixer-Oscillator ICs

Abstract

Monolithic ICs functioning as the mixer and oscillator with an impedance-matched output are described.

5.1. Introduction

As more and more functions are integrated in a monolithic form, the integration of tuner functions is of great interest. However, it has been limited so far, mainly due to the following reasons:

1. In tuners, the ultimate in high-frequency performance is utilized to meet sophisticated requirements. On the other hand, the active and passive elements employed in monolithic ICs are all designed for mass production. And, therefore, the elements cannot satisfy all the requirements of tuners.
2. The requirements of tuners often change and become more demanding as the conditions for signal reception and legal constraints become more complicated. The requirements for high-frequency semiconductor devices for tuners, such as noise figure and cross-modulation characteristics, become more sophisticated. It is easier to meet the changing demand with discrete devices rather than with monolithic ICs since much larger investments in human resources, funds, and time are required to develop new monolithic ICs.

However, the advantages of monolithic ICs, such as the reduction in size, number of components, and time for assembly, and improvement of reliability, are substantial.

5.2. First-Generation Mixer-Oscillator IC for Mechanical Tuners

The mixer and oscillator of the VHF tuner was first considered for integration because the requirements are clear-cut and relatively low frequencies are involved.

The first monolithic IC applied to a mechanical tuner for television sets is the CX-097, which was developed by a group at Sony, including the author, and employed a rather standard bipolar IC process. The IC had the functions of the mixer, oscillator, and output matching to the standard impedance of 75 ohms, and was introduced in the United States in 1973.

An ambitious way of integrating the whole function of the tuner by up-converting the signals of both the VHF and the UHF bands to the microwave range was investigated by a group at Philips.[1]

A number of investigations into integration of the tuner function were reported since.[2–9]

The functional block diagram of the mixer oscillator IC (CX097) and the equivalent circuit are shown in Figures 79 and 80, respectively. The cascade-connected transistors act as a mixer, and the stability of the oscillating frequency against the deviation of supply voltage is compensated internally by the usage of series connected diodes in the biasing circuit.

The output-matching circuit was introduced to meet the requirement of connection to the VIF circuit without adjustment regardless of the length of the cable required for connection. Thus, one could easily choose the most suitable combination of the tuner and the rest of the circuitry for specific requirements. The incorporation of additional circuits for the output was economically justifiable in monolithic integration.

The IC was mounted in a 14-pin dual-in-line package (DIP) with a few pins eliminated. The supply voltage was 18 volts and the operating current was 25 mA.

The frequency stability against the supply voltage and ambient temperature are plotted in Figures 81 and 82, respectively. The dependence on supply voltage is in the opposite direction for the VHF low and high bands. However, the IC was designed so that

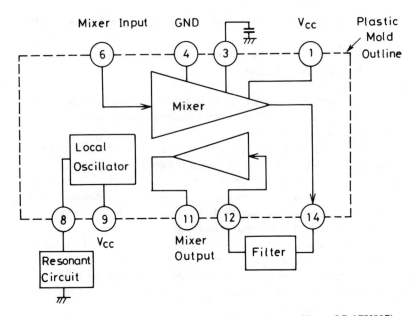

Figure 79. Functional block diagram of the mixer-oscillator IC (CX097).

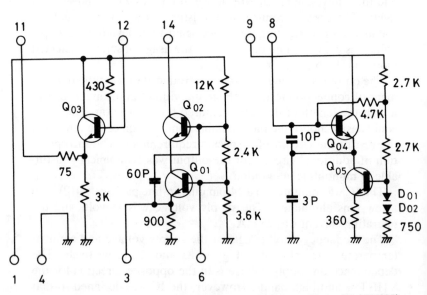

Figure 80. Equivalent circuit diagram of the mixer-oscillator IC (CX097).

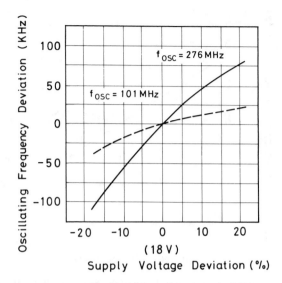

Figure 81. Oscillator frequency stability of the mixer-oscillator IC (CX097) against the supply-voltage deviation.

frequency changes caused by voltage fluctuations of ±10% stayed within 50 Hz.

5.3. Mixer-Oscillator IC for Electronic Tuners

As electronic tuners were introduced, improved performance of the basic discrete transistors was required to compensate the loss caused by the voltage-variable capacitance diodes and band-switch diodes having relatively low Q values.

The first monolithic IC applied to electronic tuners was the CX-099, which was developed in 1975 mainly by the same group at Sony.[7] An advanced bipolar IC process with an arsenic-doped emitter and an ion-implanted base was developed and this IC is still in wide use today.

The small-signal h_{fe} of the element transistor at 100 MHz is plotted in Figure 83. The performance of the transistor was among the best obtainable in those days.

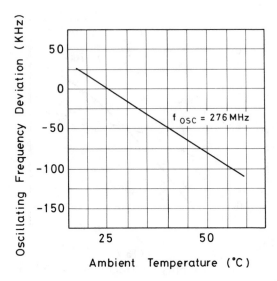

Figure 82. Oscillator frequency stability of the mixer-oscillator IC (CX097) against ambient temperature.

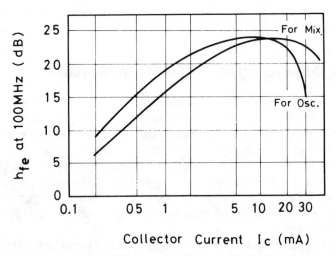

Figure 83. Small-signal h_{fe} of the element transistors measured at 100 MHz for the mixer and the oscillator.

Table 19. Electrical characteristics of the CX099

Characteristic	Condition	Min.	Typ.	Max.
Conversion gain (dB)	$f_{RF} = 200$ MHz, $f_{IF} = 43.5$ MHz	21	25	
Conversion gain (dB)	$f_{RF} = 57.5$ MHz, $f_{IF} = 43.5$ MHz	22	26	
Conversion noise figure (dB)	$f_{RF} = 200$ MHz	11.0	14.5	
UIF amplifier power gain (dB)	$f_{IF} = 43.5$ MHz	31.0	33.5	
UIF amplifier noise figure	$f_{IF} = 43.5$ MHz		3.9	5.0
Standing-wave ratio of output circuit	$f = 43.5$ MHz, 75 ohms		1.5	2.5
Starting stability of local oscillator (kHz)	$f_{osc} = 276$ MHz, 3 s–3 min		±60	±150
Temperature stability of local oscillator (kHz)	$f_{osc} = 276$ MHz, $T_a = 25 \pm 25°C$		+70 −100	+150 −200

The equivalent circuit is shown in Figure 84, where we can see that the stability against the supply voltage is again internally compensated by the use of diodes connected in series.

The employment of a zener diode is another means for stabilization against the supply voltage, but it was found that zener diodes generate considerable noise that degrades the performance of the mixer, in addition to the stabilization circuit consuming additional power.

The IC was mounted in a standard 14-pin DIP. The supply voltage was 12 volts and the operating current was 35 mA. The performance and the specifications are summarized in Table 19.

5.4. Advanced ICs

A wide-band frequency-converter IC for VHF TV tuners covering the mixer and oscillator functions was reported by S. Komatsu et al. in 1979.[2] The IC employed bipolar transistors with a f_T of 4 GHz

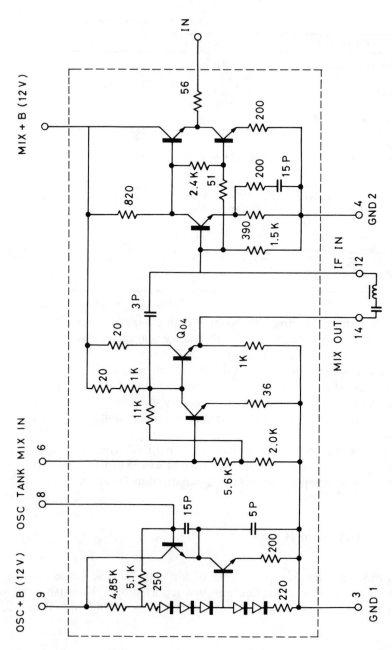

Figure 84. Equivalent circuit diagram of the mixer-oscillator IC for electronic tuners (CX099).

and covered a bandwidth of 600 MHz. The IC was manufactured by Toshiba as TA 7635P.[3]

In 1986, an IC with a similar function was manufactured by NEC as μPC 1405G.[4]

An advanced version of the CX-099 was developed by F. Ishikawa et al. in order to cover the full CATV band up to 470 MHz. It was manufactured by Sony as CX 20155, CXA1125P, and CXA1165P/M.[7-9] As an example, the specifications for the CXA1125P are shown in Table 20.

A GaAs IC for TV tuners integrating the functions of mixer and oscillator was reported by T. Nagashima et al. in 1987.[5] The IC, manufactured by Hitachi in the following year, covered the whole frequency range of VHf and UHF bands for tuners.

Although the GaAs IC has a distinct advantage in high-frequency

Table 20. Electrical characteristics of the CXA1125P

Characteristic	Condition	Min.	Typ.	Max.
Conversion gain (dB)	$f_{RF} = 470$ MHz, $f_{IF} = 43.5$ MHz	20	25	
Conversion gain (dB)	$f_{RF} = 200$ MHz, $f_{IF} = 43.5$ MHz	21	27	
Conversion gain (dB)	$f_{RF} = 55$ MHz, $f_{IF} = 43.5$ MHz	23	28	
Conversion noise figure (dB)	$f_{RF} = 470$ MHz	14.5	16.0	
Conversion noise figure (dB)	$f_{RF} = 200$ MHz	11.5	14.5	
UIF amplifier power gain (dB)	$f_{IF} = 43.5$ MHz	30	35	
UIF amplifier noise figure	$f_{IF} = 43.5$ MHz		3.0	6.5
Standing-wave ratio of output circuit	$f = 43.5$ MHz, 75 ohms		1.5	
1% cross-modulation distortion UIF (dBμ)	$f_D = 470$ MHz, $V_i = 60$ dBμ, $f_U = f_D \pm 12$ MHz		95	
1% cross-modulation distortion Mix. (dBμ)	$f_D = 200$ MHz, $V_i = 60$ dBμ, $f_U = f_D \pm 12$ MHz		94	
1% cross-modulation distortion Mix. (dBμ)	$f_D = 41.25$ MHz, $V_i = 60$ dBμ, $f_U = f_D \pm 12$ MHz		83	

operation, the productivity is still much lower than silicon. However, as the material and process technologies advance, the GaAs
devices will become widely used like silicon ICs.

5.5. Summary

Integrated circuits that have the mixer and oscillator functions with
a standard output impedance of 75 ohms were described. For ICs
for electronic tuners, advanced IC processes that provide transistors with good high-frequency characteristics are required to compensate for the relatively low Q values of the resonant circuits.

GaAs integrated circuits are beginning to be used. However,
they are not very widely used yet. Advancements in material and
process technologies are required.

References

1. An Integrated Channel Selector for VHF/UHF. *Microelectronics and Reliability* **13** (1976): 501–502.
2. S. Komatsu, K. Torii, S. Shimizu, S. Fujimori, C. Kato, K. Fujita, and H. Sawazaki. Wideband Frequency Converter IC for VHF TV Tuners, in *Digest of 1979 IEEE Solid-State Circuit Conference*, p. 236 (1979).
3. Toshiba Integrated Circuit Technical Data for TA7635P (1979).
4. NEC Integrated Circuit Application Note for μPC1405G (1986).
5. T. Nagashima, H. Mizukami, K. Sakuta, M. Shinagawa, and K. Sakamoto. A GaAs IC for TV Tuner. *IECE Tech. Rep.* **ED87-155** (1987): 19–24.
6. Sony Integrated Circuit Data for CX-099.
7. Sony Integrated Circuit Data for CX-20155.
8. Sony Integrated Circuit Data for CX-1125P.
9. Sony Integrated Circuit Data for CX-1165P/M.

Index

Japanese Technology Reviews